SPACE AND SENSE

This book reverses the usual focus on vision, and examines what active touch may have in common with vision, in order to clarify the role of the senses in spatial processing. The approach centres on the specifically "spatial" aspects of processing that are rarely specified explicitly. *Space and Sense* shows that in touch, as in vision, external as well as body-centred reference cues improve perception.

Susanna Millar presents new evidence on the role of spatial cues in touch and movement both with and without vision. She applies the notion of inter-related neural networks in current neuroscience to interactions of touch and movement with vision. The book addresses the paradox that vision is either viewed as necessary or else irrelevant to spatial processing, and challenges the traditional belief that links touch solely to body-centred information. The book shows how posture and external stimuli can bias the direction in which we move, how shape illusions in touch and vision can be reduced by the same information, and that external cues can improve location recall in touch and movement also without vision.

Space and Sense provides empirical evidence for an important distinction between spatial vision and vision that excludes spatial cues in relation to touch. This important new volume extends previous descriptions and provides new evidence for how we make sense of the space around us and what vision and touch contribute to that process.

Susanna Millar is a Senior Research Scientist at the Department of Experimental Psychology, University of Oxford.

ESSAYS IN COGNITIVE PSYCHOLOGY

North American Editors:
Henry L. Roediger, III, *Washington University in St. Louis*
James R. Pomerantz, *Rice University*

European Editors:
Alan Baddeley, *University of York*
Vicki Bruce, *University of Edinburgh*
Jonathan Grainger, *Université de Provence*

Essays in Cognitive Psychology is designed to meet the need for rapid publication of brief volumes in cognitive psychology. Primary topics include perception, movement and action, attention, memory, mental representation, language and problem solving. Furthermore, the series seeks to define cognitive psychology in its broadest sense, encompassing all topics either informed by, or informing, the study of mental processes. As such, it covers a wide range of subjects including computational approaches to cognition, cognitive neuroscience, social cognition, and cognitive development, as well as areas more traditionally defined as cognitive psychology. Each volume in the series makes a conceptual contribution to the topic by reviewing and synthesizing the existing research literature, by advancing theory in the area, or by some combination of these missions. The principal aim is that authors provide an overview of their own highly successful research programme in an area. Volumes also include an assessment of current knowledge and identification of possible future trends in research. Each book is a self-contained unit supplying the advanced reader with a well-structured review of the work described and evaluated.

Titles in preparation

Gernsbacher: *Suppression and Enhancement in Language Comprehension*
Park: *Cognition and Aging*
Mulligan: *Implicit Memory*
Surprenant & Neath: *Principles of Memory*
Brown: *Tip-of-the-Tongue Phenomenon*

Recently published

Evans: *Hypothetical Thinking: Dual Processes in Reasoning and Judgement*
Gallo: *Associative Illusions of Memory: False Memory Research in DRM and Related Tasks*
Cowan: *Working Memory Capacity*
McNamara: *Semantic Priming: Perspectives from Memory and Word Recognition*
Brown: *The Déjà Vu Experience*
Coventry & Garrod: *Seeing, Saying, and Acting: The Psychological Semantics of Spatial Prepositions*
Robertson: *Space, Objects, Minds, and Brains*
Cornoldi & Vecchi: *Visuo-spatial Working Memory and Individual Differences*
Sternberg et al.: *The Creativity Conundrum: A Propulsion Model of Kinds of Creative Contributions*
Poletiek: *Hypothesis-testing Behaviour*
Garnham: *Mental Models and the Interpretation of Anaphora*

For continually updated information about published and forthcoming titles in the Essays in Cognitive Psychology series, please visit: **www.psypress.com/essays**

Space and Sense

Susanna Millar

LIBRARY (714) 338-6215
2501 W. SUNFLOWER AVE SANTA ANA, CA 9270?

Psychology Press
Taylor & Francis Group

HOVE AND NEW YORK

First published 2008 by Psychology Press
27 Church Road, Hove, East Sussex BN3 2FA

Simultaneously published in the USA and Canada
by Psychology Press
270 Madison Avenue, New York NY 10016

Psychology Press is an imprint of the Taylor & Francis Group, an Informa business

© 2008 Psychology Press

Typeset in Times by RefineCatch Limited, Bungay, Suffolk
Printed and bound in Great Britain by TJ International Ltd, Padstow, Cornwall
Cover design by Lisa Dynan

All rights reserved. No part of this book may be reprinted or reproduced or utilized in any form or by any electronic, mechanical, or other means, now known or hereafter invented, including photocopying and recording, or in any information storage or retrieval system, without permission in writing from the publishers.

This publication has been produced with paper manufactured to strict environmental standards and with pulp derived from sustainable forests.

British Library Cataloguing in Publication Data
A catalogue record for this book is available from the British Library

Library of Congress Cataloging in Publication Data
Millar, Susanna.
  Space and sense / Susanna Millar.
    p. cm.
  Includes bibliographical references and index.
ISBN 978–1–84169–525–9
1. Space perception. 2. Visual perception. 3. Touch. I. Title.
BF469.M55 2008
152.1′82 – dc22   2007038211

Series ISSN: 0959-4779

ISBN: 978–1–84169–525–9

For Fergus

Contents

Acknowledgements   ix

Introduction: Overview, definitions and layout of the book   1

1   Concepts of space and the senses: A brief historical perspective   7
    From Socrates to Weber and Wundt   7
    Touch and vision in the twentieth century   10
    Some legacies and implications   22

2   Spatial coding as integrative processing of inputs from vision, touch and movement: The reference hypothesis   27
    Interrelations between vision, touch and movement   27
    Neuropsychological evidence and crossmodal effects   31
    Vision and blindness   34
    Experiments with blind children   37
    Cerebral plasticity and blindness   41
    Theoretical assumptions: Integrative processing of inputs from diverse sources   42
    What is specifically spatial about spatial information?   43
    Questions raised by the proposed model   45

3   Reference cues in large-scale space   47
    Initial sight or sound of a target or goal   47
    Reference information in urban environments   49
    Locomotion in large-scale spaces without stable reference cues   51
    A method and results on veering by blind children   54
    Spatial knowledge, geometric inference and perceptual cues   68

viii CONTENTS

4 **Hand movements and spatial reference in shapes and small-scale space** 71
   Finger movements and coding braille characters as shapes 72
   Spatial coding and lateral scanning movements in text reading 77
   Which hand is best? Spatial coding with the left and right hand 86
   Hand movements and spatial reference for larger non-verbal displays 88
   Haptic spatial coding and hand effects: A test of two hypotheses 89
   Summary: Spatial coding and movement information in haptic perception 96

5 **External and body-centred reference in haptic memory** 99
   Body-centred and external reference and modality systems 100
   Neurophysiological evidence for body-centred and externally based reference 104
   Indications from observation 106
   Testing external and body-centred reference in haptic tasks 108
   Practical and theoretical implications of the findings discussed in this chapter 114

6 **Visual illusions that occur in touch** 119
   The vertical–horizontal illusion 121
   Testing the radial/tangential movement hypothesis 125
   Illusions in shapes consisting of bisecting and bisected lines 132
   Conclusions 139

7 **Müller-Lyer shapes** 143
   Previous evidence on the Müller-Lyer illusion 145
   Movement time and distinctive features in the haptic illusion 150
   Shape-based discrepancies in length or size and added reference information 154
   Chance or spatial processing of sensory inputs? 158

8 **What does vision contribute to touch?** 161
   Stimulus–response compatibility and irrelevant spatial cues 162
   When and why is spatial vision "noninformative"? 163
   What aspects of vision facilitate haptic performance? 170
   Summary and conclusions 178

9 **How far have we got? Where are we going?** 181
   Moving through large-scale space: Inferences from veering 182
   Finger movements and spatial coding in small-scale space 182
   External cues can be used for reference in haptic tasks 186
   Common factors in illusions by touch and vision 188
   Vision improves haptic recall of locations only if it affords additional reference cues 189
   Conclusions and outlook 190

References 195

Author index 221

Subject index 229

Acknowledgements

I should like to thank the Department of Experimental Psychology of the University of Oxford, the head of Department, Professor Oliver Braddick, and my other colleagues who provide the stimulating intellectual climate that characterises this Department. I am particularly grateful to Professor Braddick for his support, and for further enhancing an atmosphere in which independent work can flourish.

My gratitude is due to the Economical and Social Research Council who supported my research over many years, and to the Guide Dogs for the Blind Association for a research grant that supported a number of the recent studies that are reviewed in this book. My editors, Tara Stebnicky and Sara Gibbons, have been most helpful in keeping me up to the mark. I am indebted to Professor Michael Tobin, Professor Morton Heller and Professor Alan Baddeley for very useful comments. Above all, I want to thank my former colleague, Dr. Zainab Al-Attar, for carrying out a great deal of the experimental work and analyses with a dedication, unwavering cheerfulness and sympathy that I appreciate and miss.

The book is dedicated to my husband whose love and encouragement made it possible to write this book.

Introduction

Overview, definitions and layout of the book

How do we make sense of space? What do the senses contribute? How do they relate to each other and to concepts of space? These questions go back well over two thousand years and are crucial issues in current cognitive neuroscience. They are central to this book.

The reason for the book is exciting new evidence on how a seemingly most unpromising sense modality contributes to "making sense" of space. Vision is undoubtedly the sense that is most closely associated with space perception. Most studies focus on vision. The two are even sometimes identified. But it is precisely because vision is so closely identified with spatial processing that it has often been difficult to separate what is specifically "spatial" from sensory aspects of processing information. That emphasis is almost reversed in this book. Touch is used as the starting point, because that apparently unlikely spatial source makes it potentially easier to pinpoint how spatial processing takes place and how the senses – and particularly touch and vision – relate to each other and to that process.

I have an historical alibi in foregrounding touch. In a rare treatise on touch, Katz (1925) not only deplored the usual neglect of touch in psychology, but also stated his conviction that effects of deprivation of the skin senses would alter perceptual theories more than evidence from lack of vision or hearing. Nevertheless, almost an apology seems to be needed for using touch more than sight or hearing as the springboard for considering how the senses relate to each other and to spatial processing. Hearing, as the other major source of "distal" stimuli, is certainly important in blind conditions. The direction and location of an external sound source can be gauged from time differences in the arrival of sounds at the two ears. Variations in the intensity of a sound can convey variations in the distance of a moving external object. Large obstacles can be detected by echolocation. Clapping your hands or stamping your feet produces sounds that echo from a near wall and disclose its solidity or hollowness, or a gap when there is no echo. It is even possible to use sounds to describe the shape of a shadow cast by an object. A simple light-sensing gadget that emits contrasting tones for dark and light can be used to describe the shape to a blind person by moving the device repeatedly across the line that divides the shape of the shadow from the lighter surround. Known sounds are useful reference cues for orienting movements in familiar environments. However, unfamiliar sounds do not provide reliable external references or guides in new environments when vision is excluded.

Contrary to what is still the most common notion, touch can be a crucial source of spatial information. Touch depends on direct contact with the source of stimulation and provides concretely felt feedback from obstacles. But the importance of cues that can specify the location of an object or target is more obvious when touch involves exploring and scanning movements. The combination of inputs from touch and movement – usually termed "haptic" perception – tends to be more efficient than a passively felt touch on the skin. Moreover, in the dark, when you come back to sit at your desk, you can pick up your pen without scattering all the other items on the desk, if you know where the pen was relative to your body midline or some other body-part or posture when you were sitting at the desk before, so that you can reach for it directly.

Such deliberate strategies may seem unnecessary in vision. Objects are normally seen relative to other objects or to surrounds. Nevertheless, touch and vision are closely connected in function. Shapes and spatial features are perceived by touch as well as by vision. It has also long been evident from crossmodal studies with monkeys and human infants, as well as with human adults, that shape and spatial information from vision and touch transfers across the two modalities.

The crucial role of reference cues for coding inputs spatially first became obvious to me from findings with congenitally totally blind children. The children tended to be accurate in spatial tasks if the location of an object, or its distance, could be remembered relative to the body midline when sitting upright, or to some other body-part. Disrupting that relation experimentally produced predictable errors in matching, recognition and recall. At the same time, there was evidence that total blindness did not preclude adequate and even superior spatial performance. I explained the apparent discrepancy in findings by the type of information that is available in different task conditions. It can come from vision. But it is also available, or can be made available to an individual, from other sources. However, the explanation made it important to specify precisely what is to count as "spatial" in a spatial task, or spatial performance.

Criteria for spatial coding are rarely made explicit, possibly because it has been considered unnecessary for visual tasks. The need for explicit criteria for spatial coding that can be identified and manipulated experimentally was suggested by studies without vision. It centred on the type of reference information that is available to specify where an object is, or how far it is from another object or from the perceiver.

The importance of reference information for spatial coding is not really in doubt. The problem is that the metaphor "frames of reference" is commonly used to refer to such information. It brings to mind concrete frames that surround pictures, fences around fields, or better still, the square surrounds of maps with markings along the vertical and horizontal sides that divide them into useful orthogonal coordinates. The use of the term "frames of reference" as a description of how people represent spatial information mentally is thus highly controversial. The problem is considered further in Chapter 2. But I want to make it quite clear from the outset that the term is not used in that sense in this book. The assumption throughout the book is that spatial processes are activities of the organism that integrate inputs from diverse sources to act as reference cues.

However, the term "frames of reference" cannot be avoided altogether. A traditional notion had been that frames of reference differ with modality. Vision is associated with "allocentric" or external "reference frames". They are sometimes identified with mental spatial representation in terms of Euclidean geometric axes. Haptic inputs and movements are traditionally associated with "egocentric" or body-centred "frames of reference". They also tend to be identified with mental representations. But they are often considered to be less "veridical", or more prone to distortions. The assumptions are discussed critically in Chapter 2 and later.

At the same time, the term "frames of reference" can be useful in so far as it refers to the broad classification of inputs by their source either in the external environment, from objects, or from within the organism. But the classification is not, as such, a criterion for what is to count as spatial processing.

Spatial coding is defined explicitly here as the activity of integrating inputs from diverse sources as potential reference cues that specify the location ("where?"), distance ("how far?") or direction ("what turnings?") in perception and action that a task demands. The criterion that I adopted, and which is used here, makes no *a priori* assumptions about forms of mental representation. It is an operational definition. Potential reference cues from different sources can be varied experimentally and tested for their effects on behaviour.

The theoretical description on which the criterion was based used the notion of activities in interrelated networks for the behavioural findings. The analogy was with descriptions of neural activities that were beginning to replace traditional models of purely linear processing from sense receptors to higher centres in the nervous system and brain. Network descriptions also fitted the behavioural findings better than previous rather rigid architectural models of linear progressions from perception to thinking. The metaphor has the advantage that the same terms are applicable to both psychological and neurophysiological domains of enquiry.

Chapter 2 describes the findings and methodological considerations that prompted the theoretical description that I adopted. It explains the assumptions on which the criterion for spatial coding was based and the further questions that derived from the theoretical description that are central to this book.

One implication was that the form and accuracy of spatial processing depends on the availability and reliability of inputs from all sources that can potentially be integrated to serve as reference cues. It implied that people use any available external cues also in environments – such as total blackouts or dense fog – that conspicuously lack reliable convergent inputs that can be integrated reliably. The prediction was that in such contexts, isolated cues mislead rather than guide locomotion.

The contribution of active scanning movements to spatial aspects of performance is examined with findings on recognising single braille characters as compared to scanning movements in continuous texts. It led to questions about what we can infer from the laterality of hand movements in spatial tasks. Evidence for the prediction that the efficiency of spatial coding is increased by the redundancy of compatible cues from different sources was also needed.

Another conjecture that needed testing was that discrepancies in spatial reference cues can produce errors and perceptual bias. The question was how far these may explain similarities, or show a common basis, for "optical" illusions that also occur in touch. The further question was what aspects of vision improve haptic performance and why. The studies that were based on these questions form the background to the remaining chapters.

The new findings were exciting, because they support many of the assumptions in the original description. But they also go beyond it. The point here is to consider the new results carefully in the wider context of other evidence on the basic questions of how spatial processing takes place and how the senses, particularly touch and vision, relate to the process.

Chapter 1 traces the historical origins of some relevant theoretical controversies that still colour current debates. It shows that apparently incompatible assumptions often rest on different, but equally relevant, grounds, suggesting that adequate explanations need to embrace both. The "nature versus nurture" controversy is one such misleading dichotomy. The review may also help to clarify the multiple diverse meanings of some handy abstract terms.

The meanings are rarely questioned, because the terms derive from ordinary language. But their use in psychology and cognitive neuroscience is not necessarily identical or unambiguous.

The starting point for considering effects of external and body-centred cues in large-scale space in Chapter 3 is the well-known confusion of directions produced by dense fog and total darkness. These conditions make people diverge or "veer" from the straight-ahead path or facing direction. Sounds and cues from body-postures that normally help to maintain the direction of walking mislead when they occur in isolation.

Chapter 4 focuses on the function of exploring and scanning movements in small-scale patterns and spatial layouts in tabletop space. Both have important practical uses in blind conditions. They are also essential for understanding how spatial information is processed by touch and movement. Implications of findings with the tiny dot patterns of the braille script are considered first. The functions of the left and right hands in spatial tasks that involve touch and movement are examined in the context of instructions to use reference cues in the recall of locations.

The reference hypothesis is taken a step further in Chapter 5. The chapter focuses on exciting new evidence that deliberately provided concrete external frames around a target, or a series of targets, can be used for reference also in purely haptic conditions. It also discusses findings which show that the combination of external and reliable body-centred cues produces twice the level of accuracy in location tasks than either external or body-centred cues alone.

"Visual" illusions that occur in touch are discussed in Chapters 6 and 7. These shape illusions raise intriguing questions that go to the heart of the issues considered here. The chapters examine findings which suggest that modality-specific effects are involved, but also produce evidence on factors that the illusions have in common.

The role of vision in haptic tasks is the main topic of Chapter 8. The functions and limits of the contribution that vision makes to haptic tasks are considered in relation to the type of visual information that is available in different experimental and task conditions. It is shown that specific aspects of vision do not combine in precisely the same respects with inputs from touch and movement. Moreover, added vision that excludes all spatial cues does not contribute to the accuracy of haptic memory for locations, or to error patterns for different locations. The findings suggest that vision enhances haptic perception and memory in spatial tasks only to the extent that it contributes spatially relevant information. It is argued that spatial and non-spatial vision need to be distinguished explicitly in theories of sensory integration and in bimodal studies.

The final chapter summarises the findings and implications and provides an updated description of how touch and vision relate to each other and to spatial processes, and the further questions that these raise.

CHAPTER ONE

Concepts of space and the senses: A brief historical perspective

The question of how the senses relate to ideas of space goes back at least to the beginnings of Western thought. A brief dip into its history traces the ideas that have left an imprint and how they changed with new approaches and evidence. It may also serve as a reminder of how the meanings of words that have come into psychology from ordinary language become extended with new technologies that provide useful, fresh, but sometimes confusing, analogies. The chapter is intended as an overview of the changes in how the question has been viewed over time, and why it continues to be an important issue in current cognitive neuroscience.

FROM SOCRATES TO WEBER AND WUNDT

The idea that spatial concepts are known before birth was taught by the Athenian philosopher Socrates some two and a half thousand years ago. It is nicely paradoxical that we really ought to credit that arch "rationalist" philosopher with the first empirical study of spatial reasoning. Unfortunately, an Athenian court condemned Socrates to poison himself with hemlock for teaching philosophy to the young. So we can only go by the writings, as construed by later scholars, of his friend and student Plato (c. 429–347 BC). According to Plato, Socrates' questions elicited perfectly correct geometric deductions from an untaught, ignorant slave boy. As any editor of a self-respecting scientific journal would, no doubt, have pointed out, the finding is flawed as empirical evidence for an innate concept of space. Socrates would really have to alter his method, and use better controls for leading questions and graphic demonstrations drawn in the sand, if he wanted it published. Socrates actually anticipated the objection and refuted it, at least to Plato's satisfaction. The view that concepts of space are innate has been with us, in some form, ever since.

The German philosopher Immanuel Kant (1724–1804) is usually classified as a "rationalist" philosopher. The contrast is with "empiricist" philosophers (see later), who assumed that all concepts are derived from experience. In Kant's important (1781) work, space and time are described as *a priori* intuitions. They are necessary forms in which objects appear to us. His views are sometimes cited in support of the notion that concepts of space are innate. In fact, Kant's meticulous analyses of *a priori* concepts are "a critique of pure reason" that says nothing remotely as simple as that.

Darwin's theory of evolution (1879) provided the biological framework that links the structures, functions and adaptations of organisms to their ecological niche. The enormous advances in modern genetics and microbiology are beginning to unravel the very complex interrelations that are being revealed. The interrelations being discovered at micro-biological and macro-biological levels are not, of course, to be confused with the notion of innate concepts. The assumption of an innate concept of space raises more questions than it solves about the bases and role of logical inferences in spatial perception and in understanding geometry. But so far, it is easier to quote than to question the malleability of paths that lead from genes to geometry.

The British "empiricist" philosophers – Locke (1632–1704), Berkeley (1685–1753), and Hume (1711–1776) – made diametrically opposite assumptions about the origin of concepts than "rationalist" philosophers. All knowledge derives from discrete sense impressions and contiguous associations between them. The notion that everything is learned by association underlay many of the subsequent "associationist" psychological theories in the English-speaking world in the nineteenth century. But the idea that perception is learned led to a different set of conundrums. The eponymous question that Molyneux put in a letter to his philosopher friend, John Locke, exemplifies one of these. Would a man, born blind, but who regained vision later, recognise by sight alone objects and shapes that he had previously known only through touch?

Locke and Molyneux undoubtedly expected a negative answer. It was given over two hundred years later by von Senden (1932). Von Senden reviewed the outcome of cataract operations on the eyes of people who had been totally blind from birth. The operation failed to restore visual shape perception. But his findings do not actually support empiricist assumptions. To anticipate, much later studies showed that the total absence of visual stimulation prevents the normal development of the mechanisms of visual perception (Adams, Bodis-Wollner, Enoch, Jeannerod, & Mitchell, 1990;

Chow, Riesen, & Newell, 1957; Novikova, 1973; Riesen, 1947). But more recent findings also show that spatial inference tasks are not necessarily worse for blind adults than for the sighted (e.g. Hollins, 1989).

The seeming paradox is important, and it is discussed in detail in the next chapter. Here it is sufficient to note that apparently paradoxical findings are usually evidence that an underlying dichotomy needs to be discarded. The dichotomy that underlies the view that vision must either be essential for spatial perception, or else irrelevant to it, will not do. The caveat, already cited by Katz (1925), that task performance should not be confused with conjectures about underlying mental strategies also needs to be heeded. Molyneux's question is still with us. But we are nearer unpacking the complex factors that the question actually involves.

The point is that legacies from initially useful, but nevertheless mistaken, dichotomies often linger on in new explanations. Even a cursory look at subsequent theories of perception still shows effects of previous dichotomies, such as opposing "innate" and "acquired" bases of perception and thinking. In fact, many apparently contradictory theories are actually based on findings that are relevant to subtly different questions and depend on different methods.

A line of evidence that was sometimes blamed later for producing "atomist" theories of perception can be traced to physiological studies of the specialised sense receptors and their neural connections that took off in the eighteenth and nineteenth centuries, especially in Germany. In fact, the studies were based on a search for quantitative laws that would describe how physical stimuli that impinge on an organism relate to what is sensed or felt. The most prominent protagonists were Weber (1796–1878) and Fechner (1801–1887). Weber was the originator of Weber's law on sensory thresholds. It still holds, though only for a restricted range of stimuli. Fechner took this work further. Both men had at least one foot in the realm of physics of the time. Their work originated the psychophysical methods that were central to progress in experimental psychology and also laid the bases for later computational descriptions.

Of particular interest in the context of the present book are Weber's (1834 & 1846/1978) treatises on the tactile senses and his conclusions on spatial effects in touch. Weber's interests ranged widely over the central nervous system and the structure and function of the special senses. The treatises reported his extensive observations and empirical studies on touch. Variants of the two-point threshold technique are still often used to measure tactile acuity. The threshold is the smallest distance between the two points of a protractor touching the skin that is needed for them to be felt as two separate points rather than as a single touch. Weber showed that tactual acuity of the skin differs at different parts of the body. He experimented with the skin's sensitivity to gentle touch, temperature, pain, and pressure and their relative contribution to perception, at a time when little was known about the skin receptors that such stimuli excite. He also experimented with the factors that contribute to passive and active judgements of weights.

Weber assumed that we are conscious of separate sensations, or mosaics of sensations. But the charge of "atomism" hardly fits his view. Weber was convinced that we are innately constituted to interpret sensations in terms of the categories of space, time and number. According to him, representation and memory is impossible without such categorisation, although we are not conscious of it. It has sometimes been suggested that he was influenced to some extent by the "*Natur-Philosophie*" that was current in Germany after Kant. However, Weber was a physiologist as regards explanations of both structures and functions. He argued that sensations of pressure and resistance to force, due to structures in the skin, were important in our assigning sensations to external objects, rather than to sensations arising within the body. He also emphasised the importance of scanning movements in touch. In his view exploring the contours of a large object enables us to construct an idea of its shape. The view did imply a clear distinction between felt sensations and perception that, according to him, inevitably depends on unconsciously interpreting impressions in terms of space, time and number. He attributed these unconscious interpretative activities to all "higher" animals. Cats and dogs respond differently to external objects that impinge on their senses than they do to internal sensations that make them sniff or scratch their own limbs.

Interestingly enough, one of the most celebrated of the nineteenth-century physiologists, von Helmholtz (1867, 1896), proposed that perception is mediated by "unconscious inference". The assumption that perception involves cognitive or higher-order quasi-cognitive factors of which people are not necessarily aware survived in a number of later perceptual theories and was vigorously opposed in others.

The establishment of psychology as an independent science can be traced to Wundt (1832–1920), who instituted the first psychological laboratory in 1878. Wundt (1862, 1898) emphasised the importance of experimental methods in psychology. But he also maintained that systematic analyses of conscious experience are appropriate tools for psychology. He concluded, from introspection, that thinking depends on abstract concepts or imageless "conscious attitudes" (Watt, 1905). The view contrasted with the conclusion by the American psychologist Titchener (1909) that we mentally represent objects and meaning in (typically visual) images.

TOUCH AND VISION IN THE TWENTIETH CENTURY

Many of the "schools of psychology" in the early part of the twentieth century, including behaviourism, concerned theories of learning. Learning to associate repeatedly presented separate items, or Pavlov's (1927) methods of conditioning, were basic paradigms in different accounts.

The "Gestalt" school proposed an influential theory of perception. It maintained that what people actually see are global configurations ("*Gestalten*"), not separate sensory stimuli (Koffka, 1935). We perceive whole shapes from the outset. Perception of global configurations is primary and unlearned. Empirical demonstrations showed that visual perception of a shape can change without any change in the stimulus inputs. Ambiguous figures are examples of shapes in which the same stimulus information is perceived differently in the course of viewing it: A shape that looks like a staircase changes into an overhanging cornice; a configuration that is seen as a vase on a dark background changes into two dark profiles facing each other in the foreground against a light background. The Gestalt-school laws of perception detail characteristics that produce coherent configurations, such as the proximity of items to each other or the continuity of motion of moving items (Wertheimer, 1912/1932).

David Katz (1884–1953) is considered next as one of the few major figures in the history of psychology who paid as much, or even more, attention to touch than to visual perception. In the preface to his major work on touch (1925), Katz argued that psychologists had neglected touch unreasonably in favour of what were considered the "higher" senses, particularly vision. Katz's strictures about the neglect of studies on touch seem to be aimed mainly at the contemporary Gestalt school of psychology. He made the point that, because their empirical studies of perception were based on vision, the Gestalt school laws of perception placed too much emphasis on the primacy of shape and form at the expense of other spatial and temporal aspects of perception. Katz actually agreed with much of the Gestalt-school approach to the study of perception. Like the Gestalt school, he explicitly rejected "atomistic" sensory theories and the notion that perception is mediated by unconscious cognitive inferences. He also argued explicitly for a phenomenological approach, which analyses the perceptual phenomena that people actually experience. Katz himself cites the phenomenology of the philosopher Husserl as his major influence in that regard. However, he came to disagree with a major main tenet of the Gestalt school's theory of perception: that we perceive whole shapes from the outset. It did not accord with his findings on touch.

Phenomenological aspects of perception are no less important in touch than in vision, according to Katz. He argued that theories of perception would benefit by taking the skin senses into account instead of concentrating solely on vision and hearing. Perceptual theories would change, particularly in accounting for the effects of perceptual loss, by considering evidence on loss of perceptual inputs from the skin senses.

Katz tested perception by touch with people who had lost a limb during the First World War. Surprisingly, amputees were able to recognise objects with the remaining stump, much as they would have done with the now amputated hand, even if not quite as accurately. He also found that amputees reported that they "perceived" sensations in the now amputated or "phantom" limb, especially in missing hands. The findings are now well established and extended.

Katz considered himself to be an empirical scientist rather than a theorist. That may have contributed to his being one of the earliest psychologists to dispense with some time-honoured dichotomies that divided previous and many subsequent theories. He considered the "innate/learned" controversy useless. He also avoided descriptions in terms of high/low, global-unitary/local-separate categories as if they constituted irreconcilable alternatives. The very title of his major work on touch, *Der Aufbau der Tastwelt* (1925), attests to his agnostic attitude to rigid categories. The German title includes the word "*Aufbau*", which literally means "build-up" or "construction". That term is usually omitted in the translation of the title (*The World of Touch*, 1925/1989), presumably to avoid any indication of the kind of "sensory atomism" that Katz detested. Nevertheless, Katz used the German term for construction to describe the relation between simple and more complex tactual phenomena. He also used the notion of an "elementary formative" factor to designate the crucial role of movement in the build-up of complex tactual perception.

Katz already emphasised the multisensory character of touch perception, although he did not use that term. But he suggested that the perception of objects depends on a "composite impression" of diverse inputs. The diversity of sensation felt with the skin – including heat, cold, pressure, pain, hardness, softness, as well as gentle touch – is, of course, known to depend on different skin receptors (see Mountcastle, 2005). Katz also suggested that most tactual phenomena depend on movement and involve kinaesthetic and proprioceptive inputs from muscles and joints. Katz thus anticipates the notion of multisensory inputs for haptic perception by suggesting that movement, kinaesthetic and proprioceptive cues, as well inputs from the skin receptors, converge to form what he called "composite impressions" of objects.

Katz was also concerned with the relation between touch and vision. He described the tactile perception of elasticity and transparency that corresponded to colour phenomena that he had analysed earlier. More important, the hand perceives continuous surfaces, despite the gaps in touch-stimuli from separate fingers; this is analogous to the continuity in visual perception, which fails to register the "blind spot" at the entry point of the optic nerve to the retina, which lacks receptors at that point.

His resistance to theoretical absolutes, as well as his findings on the complexity of perceptual phenomena in touch, led Katz to emphasise the composite character of perception. It anticipates the notion of modalities as sensory systems (Gibson, 1966). Moreover, Katz, like Weber, distinguished between tactual impressions that we assign to external objects and those that we experience as subjective. Tactual phenomena are essentially bipolar, according to him. The subjective pole predominates for pain and in passive contacts that are not habitually used to identify objects. According to Katz, movements of the "touch organ" – mainly the hand and fingers – favours objectification relative to an object and feeds into the "spatial sense of the skin" (Katz, 1925).

It is not entirely clear what Katz meant by a "spatial sense" or how it relates to vision. Katz stressed that sighted people often report using visual images in memory tasks and that shape is often perceived better by vision than touch. But he did not propose that spatial performance is impossible without vision. In his view, vision lacks the bipolarity of touch. Visual impressions are nearly always projected onto external space. The sighted cannot escape from visualising tactual impressions. Vision may thus underlie perception of external space, at least for the sighted.

Revesz (1950) considered it self-evident that "visual space" and "haptic space" differ and are totally autonomous perceptual phenomena. Kinaesthetic inputs and movements in touching and grasping objects create haptic space, which parallels, but is not identical with, visual space. His view was based on studying blind people's experience and performance in what are usually known as the "visual" arts. His findings suggested differences between blind and sighted people's estimations of distances and other spatial tasks. In his view, too, the primacy of shape perception suggested by the Gestalt school only applies to vision, and not to touch.

In touch, it is the structure of objects, not their form, that is primary. Active touch produces a constructive synthesis of forms by successive exploration. Spatial aspects of touch are not in doubt. But objects are related to the perceiver in the first instance, and not to each other. Moving hand and limbs relative to the body creates near or personal space. Haptic space and perceived directions are built up largely from grasping and scanning multiple objects. Touch provides actual contact with the world of objects. Haptic perception involves creative aspects and cognitive functions. Geometric similarity is not the same as phenomenological identity. The fact that visual illusions are also found in touch is fortuitous. They are quite separate perceptual phenomena.

The idea that the sensory modalities provide totally separate impressions seems to hark back to the empiricist philosophers, Locke, Berkeley and Hume (see earlier). But Berkeley also proposed that visual perception is learned through touch. Maybe his view, and possibly also that of later motor theories of perception, should be exemplified by a vigorous kick at a real obstacle to show how reality is encountered.

More seriously, overwhelming later evidence on binocular, stereoscopic and monocular visual cues to perceiving depth (e.g. Howard & Rogers, 1995), and the relative rate of maturation of relevant cortical areas, makes it easy to dismiss any notion that visual perception of depth is learned through touch and movement. By a nice historical twist, purely "visual" factors alone are not sufficient to explain visual perception of orientations. For example, the position of the eyes in the sockets, and the direction of eye movements relative to the head, are controlled by three sets of extra-ocular muscles. Perception of the vertical direction depends on head- and body-tilts relative to gravity.

Howard and Templeton's (1966) book is directly concerned with spatial behaviour. They review studies of spatial orientation in humans and other animals, which include physical, anatomical and physiological as well as behavioural factors. The authors conclude (p. 7) that human spatial behaviour is "conditioned ... by the way the human body is constructed and normally develops, and by ... the nature of the physical world, that is by environmental or ecological constraints". Their book was primarily concerned with behaviour that is determined by the angular position of the body in relation to any stable external reference system. These orientations can be defined and calculated in terms of geometric angles that a variable line, representing the axis of the body or of a body-part, forms relative to a fixed line that represents whatever aspect of the world is taken as the baseline, such as the direction of gravity.

Gravitational force produces the most important constraint. The structures of the neuromuscular system evolved to keep the body upright against the forces of gravity. The mechanical structure of each joint determines the range and direction of a limb movement. But the orienting systems of the body, relative to the force of gravity, and relative to the supporting surface, are crucial for spatial behaviour to be adaptive.

A number of postural reflexes operate at birth. But adaptation to external constraints is important. Classical and operant conditioning are mentioned as the relevant mechanism for adaptation. Spatial judgements usually involve several sensory modalities in various combinations. Inputs come from vision, from hearing, from visceral sources, and from receptors in joints and muscles that are involved in producing movements and in information from movements.

The vestibular apparatus in the inner ear is particularly important. It consists of three semicircular canals and a sensing organ that register the force of gravity, acceleration, and linear and rotary motion relative to gravitational force. Behaviourally, the information is necessary for discriminating movements and the position of the body and of body-parts and postures.

The vertical direction of the mid-body axis in the upright posture is usually kept in line with the (vertical) direction of gravity (Figure 1.1). It serves as an important reference axis. Howard and Templeton (1966) defined

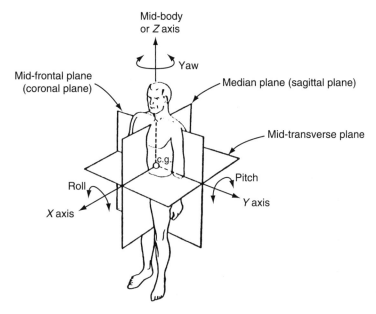

Figure 1.1 Body coordinate system and axes of rotation. [From I. Howard & W. B. Templeton (1966), *Human Spatial Orientation*. Copyright © 1966 held by John Wiley & Sons Ltd. Reproduced with permission of the author and the publisher (copyright holder).]

egocentric orientation as the position of an object, or of a limb, with respect to the body of the observer. The observer's body midline often serves as an important reference (Figure 1.1).

The visible world is described in terms of the visible surroundings that light sources and reflecting surfaces make available to an observer at a particular location. Visible surrounds are important for locomotion and for reaching to a target. The fact that different sensory modalities interact in contributing to spatial performance is evident from errors that result from conflicting cues. Plasticity in human sensorimotor coordination, and a role for corrective feedback from self-produced movements, was demonstrated in a series of studies by Held and his colleagues (e.g. Held, 1963, 1965; Held & Bauer, 1967; Held & Hein, 1963). Experiments on viewing felt targets through distorting spectacles showed that people localised the target where they saw it, and not where it was felt (e.g. Rock & Victor, 1963).

Vision tends to "dominate" tactile-kinaesthetic location cues when the two are made to conflict. It should be noted here that this is not always the case. Consider sitting in a train prior to departure. Seeing another train apparently moving when you look out of the window will make you think your train has started, until you check whether you can feel it move. Apparent discrepancies in findings led to the controversy whether visuo-spatial or tactile-kinaesthetic and movement modalities "dominate" location judgements. Howard and Templeton (1966, p. 417) make the eminently sensible suggestion that the dispute would be resolved by providing operational definitions of the frame of reference that is used in different tasks or conditions.

One of the most influential of twentieth-century theories was evolved by J. J. Gibson (1966, 1979) and E. J. Gibson (1969). They rejected the phenomenological approach of the Gestalt school. But they were also completely opposed to the notion that perception is constructed from isolated sensations. They produced a great deal of empirical evidence in support of their theory. The business of psychology is to explain how an animal or human observer perceives the information that is available in everyday life. We see people, objects and scenes, not successions of light points, nor inverted retinal images.

Gibson (1966) proposed that the sensory modalities are complex perceptual systems. They do not correspond to the popular notion of the "five senses". Single stimuli are rarely, if ever, sensed as such by the organism. The perceptual systems have evolved to detect information actively. Their function is to seek and extract information that enables the organism to orient itself, to move and to detect objects in a three-dimensional world. Perceptual systems depend on the anatomical structures and functions of the whole organism and involve neural and brain connections of the sense organs. Passive or imposed stimulation is perceived through proprioceptive information from the body. The sense organs in the systems contain a variety of receptor cells, which respond to mechanical or chemical energy or light stimulation from the environment. Stable reference-axes for the orientation of the eyes, ears, mouth and nose – that is to say, for the organs of the visual, auditory and taste-and-smell perceptual systems – also depend on postural information of the head relative to the body and to gravity (Gibson, 1966, 1979).

The human visual system evolved to detect structures in ambient light that specify the environment by differences in reflected spectra and light intensities. The eyes are highly differentiated structures. The light/dark (rod) and colour (cone) receptors in the retinas connect via long nerve fibres to a series of dedicated cortical areas. But information from overlapping optic arrays also depends on the position of the observer. That changes with body, head and eye movements, and reveals different perspectives and boundaries.

What is usually referred to as "touch" depends on a complex haptic system that involves touch and movement and includes the whole body and bodyparts, the limbs, and all of the skin surface with its multiple sensing devices that function in the perception of the body surface. It is also, literally, "in touch" with the external world. Unlike the other perceptual systems, the haptic system functions as a performative system, as well as an active perceptual system. It obtains information by exploring external surfaces and objects, especially with the hands and fingers. But these also have executive functions. They perform actions that can displace, mould, build or otherwise alter the environment. Active perception and performance involve movement. They consequently depend on the structure of the human skeleton and on how the bones are hinged at the joints. Proprioceptive information from the joints registers the angles of postures and changes of posture of the limbs, and of the head relative to the body and to the gravitational direction. In combination with kinaesthetic information from movements and tactual information from the skin, that information also normally overlaps with information from the other perceptual systems, including vision. But the haptic system can also be used without vision (Gibson, 1962).

All perceptual systems provide proprioceptive, body-centred information as well as exteroceptive information about external surfaces and objects. Children and adults also perceive what actions a surface or object "affords". Some invariants are detected from the outset. New-born infants orient to new stimuli (Wertheimer, 1961). They look longer at a new stimulus than at a familiar one that they have seen repeatedly. That provided a new technique for testing what infants see. Very young infants can discriminate distinctive shapes. From the age of six months infants refuse to crawl across a glass plate if the patterned surface below the glass provides visual depth information. The edge of the glass thus becomes a "visual cliff" (Gibson, 1969). Some experience is involved. But there is no need to retrieve that knowledge from memory prior to perception. Adults perceive directly what actions a letter box affords when they see one. Perceptual learning occurs. But it is not associative, nor does it exemplify classical or operant conditioning. It consists of progressively more detailed differentiation of distinctive features of objects and better detection of the invariant features of the environment. The modality systems pick up multiple body-centred cues concurrently. Practice makes it possible to isolate more subtle invariant aspects of the information that is potentially available to an organism (Gibson, 1969).

The business of all perceptual systems is to extract invariant information from lawful changes in spatial and temporal information that movements of the body relative to the geographical environments provide. Modality-specific inputs – such as colour in vision, sounds in hearing, or heat, cold, pressure, and touch from the skin – provide extra information. But they are relatively unimportant. Only atypical sensations lead us to notice differences between modalities. All modality systems pick up the same higher-order relational information (Gibson, 1969).

Gibson (1966, 1979) embraced the growing body of evidence from all levels of biological science that relate the structures and functions of the body to the physical world. In common with Howard and Templeton, he emphasised the orienting system of the body relative to the forces of gravity. The supporting surface is crucial. The vestibular apparatus registers the forces of gravity and interacts importantly with other organs and perceptual systems. It contributes to upright posture and equilibrium by specifying the forward, backward, right or left tilt of the body. It initiates compensatory reactions of anti-gravity muscles. Combined with the perceptual system of the skin, it contributes to postural information about the body's orientation to the ground or supporting surface. The combined systems are concerned in an implicit awareness of the up–down (head-to-foot), right–left (trunk–limbs) and front–back three-dimensional directions of the body reference frame, and of the vertical and horizontal axes of the gravitational reference frame of the world.

Space perception is the perception of the layout of surfaces in relation to each other and to the ground. The relations change as the organism moves around. Empty space cannot be perceived. Gibson (1979, p. 126) extended the suggestion (Katz, 1925) that touch perception has an objective and a subjective pole. Perception of one's body accompanies perception of the environment. The two are inseparable aspects of the same information.

Spatial perception is amodal but direct. Spatial information is out there to be perceived. It is detected or sought, not constructed or construed. Neither memory nor cognition is involved. The permanent possibilities of optical stimulation, and of moving points of convergence in a geographical environment, correspond to the set of all paths of locomotion. Changes that are contingent on moving about produce new, but overlapping, samples of optic arrays and of kinaesthetic inputs. These establish the invariant features of the environment that are detected by perception. The perceptual systems of organisms have evolved to detect such invariant and ecologically valid information (Gibson, 1966, 1979).

Gibson's exhilarating outburst (1979, p.3) was seductive: "Space is a myth, a ghost, a fiction for geometers . . . if you abandon the dogma that 'percepts without concepts are blind', as Kant put it, a deep theoretical mess, a genuine quagmire will dry up." It is unnecessary to postulate the existence of cognitive maps (Tolman, 1948) in either rats or man. It is irrelevant whether the rat runs or swims, or uses vision, touch, smell or hearing to reach its goal. Invariant amodal information about its position relative to gravity and to the environment is out there to be perceived.

The notion of "direct amodal perception" of spatial relations is somewhat confusing. The term "amodal perception" has been used to label perceptions for which part of the usual sensory input is missing. The point of calling such perception "amodal" is that there are no current sensory stimuli for what is evidently being perceived. For instance, even young infants respond as if they perceived a complete triangle although a part was obscured by another shape. We perceive a continuous line even if inputs from the receptors are discontinuous at a particular point of the retina.

But the term is also sometimes used for perception that involves – as it usually does – inputs from more than one modality. Strictly speaking, the term "multimodal perception" is more appropriate and less misleading.

There is no lack of sensory inputs in that case. Inputs from different sensory modalities contribute to the perception.

Gibson also used the phrase "amodal perception" to argue that we perceive spatial relations "directly". We certainly see objects and surfaces relative to each other and relative to surrounds. Indeed, it is difficult to see them in any other way. Gibson (1966) argued that relational variables involve the integration of different components. They are critical to all perceptual systems. There is ample and growing evidence for that formulation. But the view that we perceive spatial relations directly in an amodal fashion, in addition to perceiving modality-specific aspects of inputs like colour or texture directly, is difficult to flesh out without assuming the duality of description that the notion of direct perception was designed to replace.

Gibson (1979) circumvented the problem by rejecting the notion of "information" that had been developed in analogy with telegraphic and telephone and communication technology. The analogy had actually been useful, for instance, to describe how the physical stimulation of receptors transforms into sequences of nerve impulses that transmit the information in binary form through neurons, via a number of relay stations in the central nervous system, to specific areas of the brain. Gibson rejected such physiological and psychological analogies. Only changes at the cellular level are relevant. He foregrounded instead effects of the physical environment on the anatomical structures that evolved to enable a moving organism to detect the changes and invariant features that are available in the "ambient energy flux" (p. 263). That is the "information" which the perceptual systems detect. That form of information pick-up can be measured in terms of orienting, exploring and adjusting activities, and by other movements of the organism. Gibson's (1979) view that explanations of behaviour need to take ecological factors into account was particularly influential, as was his view that systems of perception and action develop together (Tipper, 2004).

Some theorists opposed the notion of "direct perception" of external relations as naïve philosophical realism, which assumes that the world is constituted precisely as humans perceive it. Others still considered the possibility that people's covert "heuristics" can influence how they perceive information.

Rock (1973, 1984) used an analogy with testing scientific hypotheses to describe covert cognitive activities that his findings on visual perception led him to assume. He distinguished between environmental, shape-based and egocentric (retinal in the case of vision) frames of reference that all relate to the visual system. The perceiver assigns the top, bottom and sides of a shape or object, or the direction of a line, by relating them to one or more of these frames of reference. The orientation of visual shapes or lines is inferred from such alignments. Egocentric alignments relate to the geometry of the retinal image, whether or not the perceiver is aware of that. Environmental frames of reference are based on immediately surrounding cues or the wider environment. The main problem with that description is the apparent implication that perceiving the orientation of objects depends on conscious cognitive inference.

Rock (1997) defended the notion of indirect visual perception on empirical grounds. Evidence on the effect of orientation on recognising objects and shapes is most relevant in the present context. The findings are not in doubt. Errors and/or the time it takes to recognise an object or shape increase if the orientation of the shape is changed relative to the perceiver. Changing the orientation of the perceiver relative to the object by altering the posture of the perceiver's body, or the position of his/her head, has similar effects on recognition. Small deviations from the upright of well-known familiar objects have relatively little effect. Similarly, an observer who views a luminous vertical line in the dark when his/her body is slightly tilted still sees a vertical, or almost vertical, line, despite the fact that the image on the retina at the back of the observer's eye is tilted. The perceptual system corrects slight differences in the orientation of the retinal image by taking the body-tilt into account, although the observer is usually not aware of that. Here the description is much the same in "indirect" as in "direct" perceptual theories.

However, mere knowledge about the orientation of a shape affects recognition accuracy. It implies that covert cognitive activity does play a part. For instance, it is sufficient merely to inform an observer verbally that a simple new figure is tilted for the observer to recognise it correctly. However, even familiar faces are difficult to recognise when they are presented upside down. Rock explained this in terms of the number of items in a face that would need to be restored to the usual upright direction before the face could be recognised. Moreover, rotating a relatively unfamiliar complex configuration may prevent recognition altogether if the perceiver is unaware of the rotation. Instructing the perceiver to rotate a shape mentally to its original orientation improves recognition. Such findings suggested that, in some conditions, recognising the orientation of shapes and objects may involve covert "higher order" or central activities, whether or not people are aware of them.

Terms like attention, memory, thinking and mental representation had become respectable again in the 1950s and 1960s after being considered taboo by a majority of psychologists in the early part of the twentieth century. The change came about partly because the new technologies for the transmission of information and the development of computer systems provided new analogies that promised mathematical descriptions and fostered the invention of new experimental methods.

Not all aspects of human information processing could be explained by adhering strictly to the mathematical formulae derived from the then current communication technology. However, concepts such as the transmission of information via "limited-capacity" channels, the "coding" of information so that inputs are transformed into a form in which they can be transmitted, and the notion of "redundancy of information" proved useful in describing behavioural findings, especially those produced with new experimental methods.

Studies of vigilance in the radar tracking of aeroplanes during the Second World War showed that rather elusive terms like "attention" in perception and communication could be investigated behaviourally by using operational definitions in terms of continued accuracy in tracking moving stimuli (Broadbent, 1958). Broadbent's findings suggested that incoming stimuli are filtered before entering a "limited-capacity" channel by cerebral mechanisms that selectively favour certain classes of stimuli, such as their novelty, or the length of time since they were encountered before (Broadbent, 1958). Questions about the nature of selection of stimuli to which attention is paid, and the degree of perceptual analysis that produces early perceptual selection, are still important.

New methods also allowed experimenters to go beyond testing only for long-term memory of associatively learned items and to infer how people maintain inputs in memory across brief delays. For instance, short delays between a given input and the memory test were either left unfilled, or were filled with different stimuli or with specific activities that would interfere with different means of keeping the original information in mind (Brown, 1958; Peterson & Peterson, 1959).

A neat – but probably too neat (Atkinson & Shiffrin, 1968) – model suggested a series of three memory stores. In this model, inputs initially persist very briefly in a completely unlimited sensory store, but they "drop out" unless they are quickly recoded and transmitted in verbal form to a short-term store of very limited capacity, before being stored in longer-term memory in more permanent form.

The limits of temporary or short-term memory were of particular interest for both physiological (Hebb, 1949) and behavioural (Miller, 1956) scientists. The number of separate serial items that people can remember across short delays is severely limited (Miller, 1956). Diverse studies asked whether this is because the items simply "fade" with delay, or because there is interference from other items, or because cognitive capacity is severely limited, or because temporary memory is "modality-specific".

The new methods used in the behavioural studies on immediate or short-term memory allowed experimenters to infer people's covert heuristics in coding the inputs during short delays. Impaired recall of strings of verbal items that resembled each other phonologically showed that these items were remembered by sound (Conrad, 1964, 1971). The method of interpolating different types of activities (e.g. Brooks, 1968) during short delays was greatly extended. It provided a great deal of evidence for the importance of recoding inputs phonologically. These and other findings prompted Baddeley's influential model of Working Memory (Baddeley & Hitch, 1974). Instead of a single short-term memory store, it proposed that working memory consists of three components: an attentional system acting as a central controlling executive, and two subsidiary short-term storage systems. One was a phonological loop to hold speech-based and acoustic information in a temporary store, but which would fade unless refreshed by overt or covert articulatory rehearsal. The second temporary store, termed the visuo-spatial sketchpad, was assumed to hold visuo-spatial information, which was also assumed to involve some refreshing activity (Baddeley, 1986). With additional and recent findings also on kinaesthetic inputs, the model was evolved further to incorporate these (Baddeley, 1990, 2000, 2007).

SOME LEGACIES AND IMPLICATIONS

Perfectly reasonable and useful distinctions often give rise to unsustainable dichotomies. The nature/nurture controversy has lasted in one form or another for centuries. Indeed, it is still sometimes debated as if it concerned irreconcilable alternative explanations, rather than as an initial distinction that conceals rather than reveals the complex interacting factors that influence different functions of the organism. As noted earlier, the fast pace of research in genetics and molecular biology, and in neurological techniques, has made it feasible to discard the dichotomy and the question of "how much" two separate factors contribute. Instead, it is now possible to start considering the intricacies of "how" the immediate and wider environment in which an organism exists interacts with its biochemical and physical make-up at microbiological as well as macro-biological levels (Millar, 1988a, 2005).

A similar change took place in the empiricist legacy. The postulates put forward by the mid-twentieth-century physiologist Hebb (1949) were based on purely empiricist, associationist principles. He proposed that repeated stimulation of adjacent brain cells results in a "cell assembly", so that stimulation of any cell eventually evokes all these cells together. The assumption that "reverberating circuits", analogous to electrical circuits, keep new inputs available for short periods explained temporary or short-term memory. With repeated stimulation, the cell assemblies form wider nets of neural connections via synapses that store alterations and repetitions that are important for long-term memory. Not all his speculations were borne out by later findings. But some of Hebb's (1949) proposals, such as the critical role of neural activity in the modification of synapses by pre-synaptic and post-synaptic activity, were supported and refined by later work (Bi & Poo, 2001).

Furthermore, Hebb's (1949) rules of how the brain may work to store information, coupled with analogies based on information theory and computer technology, were seminal in the development of Connectionist models. Incremental rules could be formulated mathematically (e.g. Pouget & Snyder,

2000) and simulated on computers for input–output connections mediated by concealed layers in various multilayer networks.

Connectionist models, including the influential Parallel Distributed Processing (PDP) model (Rumelhart & McLelland, 1986), were often criticised as purely empiricist. However, there seemed to be no reason, in principle, why central "top-down" and contextual effects, and the role of innate predispositions, could not be simulated in network-like models (Millar, 2005). The fast development of computer technology, following the invention by Turing of a machine that deciphered coded enemy messages in the Second World War, provided important analogies, not merely for processing sequential data, but later also for modelling parallel activities mathematically. More recent models, made possible by advances in non-linear mathematics, now include innate and contextual effects (e.g. Bi & Poo, 2001; Munakata & McClelland, 2003).

Another controversy ensued over "direct" versus "indirect" theories of perception. Much of that controversy seems to be due to differences in the method and task conditions that are used to investigate how an organism perceives the world. For instance, whether perception involves *quasi*-inferential or cognitive activity may depend on whether the perceptual task involves identifying an object or locating it among others (Norman, 2002). It is worth noting that controversies can turn on subtle differences in questions and in the methods by which they are tested.

Psychophysical methods use detection and discrimination tasks. The interest is in measuring accuracy and precision in detecting and discriminating some stimulus attribute. Accuracy depends on the (mean) size of errors. Precision is measured by the variability of error scores. In the "method of adjustment" the observer adjusts a series of stimuli to equal a standard that, measured objectively, equals the midpoint of the series of variable stimuli. Responses that exceed the standard are signed as positive errors. Adjustments that fall short of the standard are scored as negative errors. The point of subjective equality (PSE) is the (algebraic) average of signed errors for an observer. It measures the accuracy of responses. The variability of errors around the PSE measures the precision of the responses. The order of stimulus presentation (e.g "constant method", "method of limits") depends on the best means of presenting or calculating biasing effects. Scaling methods involve asking observers to make categorical judgements about the magnitude of a sensory feature.

An important method of measuring the threshold at which a stimulus can be detected was pioneered by Green and Swets (1966). It involves two measures. The sensitivity (d') with which an observer detects a stimulus is calculated by the probability of discriminating it from "noise". The stringency of the criterion (β) underlying the decision that a stimulus is present is measured by the type of error that is made. The probability of errors due to missing stimuli increases by responding only to very clear signals. Observers whose criteria are less, or not at all, stringent wrongly identify noise as positive signals. The criterion is calculated from the frequency of such "false positive errors" and errors due to "misses". Psychophysical tasks are thus tasks of stimulus detection and discrimination. One or both of the relevant stimuli are present.

Recognition tasks differ in that respect. To test the effect of orientation on recognising unfamiliar shapes, the stimulus shape has to be presented first in one orientation. It is then re-presented after it has been tilted or rotated. Recognition tasks thus involve successive presentations, with brief delays during which neither the original stimulus nor the rotated stimulus is present. Delays, however brief, require some form of short-term "storage" or coding activities that bridge the brief gap between presentations, whether or not the observer is aware of that. It would be impossible otherwise for an observer to recognise a new shape when it is presented again in its original orientation after the delay, but fail to recognise the shape if it has been rotated during the brief interval. Perception of shape orientation in such tasks can thus be described as mediated or indirect in some sense.

The main difference in using recognition tasks to test perception does not actually lie in the fact that the inputs are successive, but that they compare the observer's performance with stimuli that are designed to differ on a relevant attribute. The interest is in what the experimenter can infer from the accuracy and/or speed of performance. It has been shown, for instance, that the simultaneous presence of two comparison stimuli does not necessarily elicit immediate perception of their relative orientations. It takes time to judge whether two complex "nonsense" shapes differ in shape or only in orientation. The time to judge correctly increases roughly in proportion to the degree to which identical shapes are rotated with respect to each other (Kosslyn, 1980, 1981; Shepard & Cooper, 1982; Shepard & Feng, 1972; Shepard & Metzler, 1971). Such findings suggest that observers were engaged in some form of covert activity before responding correctly. The observer may perceive immediately that the shapes differ, or even that a specific feature in each shape is located differently. But it takes time to decide whether the configuration of the complex shapes differs, or whether the shapes differ only in orientation. Tasks of this kind clearly do not elicit an immediate or "direct" perception that the shapes differ only in their orientation. The tasks are not intended to test perceptual acuity or sensitivity in detection or discrimination.

The factors that are varied experimentally in many tasks are changes in the stimuli that the observer needs to perceive and to rectify for accurate performance For instance, objects are presented in a vertical orientation relative to an environmental surround, or relative to an observer's upright posture. The display is then rotated to a different orientation in the test condition. The observer's performance is scored in terms of the accuracy and/or speed of recognising the object in its new relative position and the variability or precision of these responses. The important statistical differences relate to performance scores between task conditions. Differences in the degree and direction of errors that relate to changes in the position of objects relative to an observer's upright posture, compared to the relative change in the position of objects relative to an external surround, allow the experimenter to infer which form of processing produced the errors, whether or not people used these forms of processing intentionally.

Such methods, often termed cognitive approaches to perception, vary perceptual conditions in tasks that require covert processing of information before responding. Differences in the accuracy and/or speed of performance in different experimental conditions allow the experimenter to infer what the observer perceived and the heuristics that are likely to underlie his or her performance.

The new behavioural methods in studies on short-term memory that took off in the middle of the last century (see earlier) were important for gaining new insights, because they made it possible to infer also people's covert heuristics in coding perceptual inputs during short delays (Baddeley & Hitch, 1974; Henry & Millar, 1991, 1993).

Empirical methods in cognitive approaches to perception, and the findings on coding inputs during brief periods of delay between presentation and response, are complementary to psychophysical methods. They are not alternatives. The methods test different, but related, questions.

Almost all the varied theories that are outlined in this chapter had some influence on my work. Many aspects of the theories about the intricacies of genetic endowments and change, and the role of modalities in space perception and in working memory, are very relevant to current studies. Space cannot be identified with geometric principles, if only because multidimensional descriptions are possible. But Socrates' slave boy had a point. Correct geometric inferences do seem obvious. To assume that everything is learned does not accord with the notion of spatial intuition as a precondition for perception. The skeletal and muscular structures, sense organs and neural organisations that enable organisms to live and move in gravitationally oriented terrestrial environments depend on genetic preconditions. However, these also involve, rather than preclude, adaptation with learning and experience.

The changes in viewpoints over time about the relation of the senses to spatial behaviour and understanding left a legacy of valid points that are still relevant to contemporary research. These, and the findings that new empirical methods made possible, suggested the theoretical description that is considered in the next chapter. It seemed to me to account for the evidence at the time. It also prompted the further questions that are central to this book.

CHAPTER TWO

Spatial coding as integrative processing of inputs from vision, touch and movement

The reference hypothesis

The description of spatial coding examined in this chapter was influenced to some extent by most of the theories outlined in the previous chapter that seemed to me to make valid, but not necessarily incompatible, distinctions. It was prompted more specifically by two types of behavioural findings, together with the fact that they accorded better with the neurological evidence that was beginning to be described in terms of neural networks than with previous more rigid accounts of brain functions.

One was the growing evidence, from the second half of the twentieth century onwards, for interrelations between the sense modalities, including vision and touch. Behavioural findings from crossmodal studies were not in accord with the assumption of purely specialised modular systems. But the idea that modality-specific cues are relatively unimportant (e.g. Jastrow, 1886) did not accommodate all the findings either.

The other arose from the discrepancies in findings on spatial performance in blindness. This chapter discusses the findings that suggested the view adopted here and the further questions and studies they prompted that are central to this book.

INTERRELATIONS BETWEEN VISION, TOUCH AND MOVEMENT

Interactions between vision, touch and movement were demonstrated by evidence from conflicting cues (e.g. Rock & Victor, 1964; Rorden, Heutink,

Greenfield, & Robertson, 1999) and by the re-afference studies (Held & Bauer, 1967), reviewed by Howard and Templeton (1966). The use of lenses that distorted people's view of the location of objects, including the position of their own hand, showed that the visual position often "captures" or "dominates" touch in these conditions. But crossmodal studies also showed that vision and touch interact.

Crossmodal paradigms were used increasingly also in developmental studies. In these designs, a stimulus is presented in one modality, and people have to pick it out, match it or reproduce it another modality. Input and test modalities are reversed in counterbalanced order across trials. Speed and/or accuracy in crossmodal conditions are compared with scores in intramodal conditions for both the contributing modalities.

Apes, preverbal infants, nursery school children as well as human adults were shown to match information across vision and touch (e.g. Bryant, Jones, Claxton, & Perkins, 1972; Davenport & Rogers, 1970; Krauthammer, 1968; Meltzoff & Borton, 1979; Millar, 1971, 1972a, 1972b; Rudel & Teuber, 1964; Streri, 1987, 1993; Streri & Gentaz, 2003).

Even newborns orient visually to the location of sounds (Castillo & Butterworth, 1981; McGurk, Turnure, & Creighton, 1977; Wertheimer, 1961). Visual-pursuit eye movements in newborns are affected by head postures in the prone position. Visual inputs can also modify head posture (Jouen, 1992). Input–output relations thus seem to be bidirectional from the start. The brain areas concerned with vision, touch and movement develop at different rates in early infancy. Input–output relations between these modalities vary accordingly in the first year of life and with developments in hand–eye coordination, especially around the age of about four months (Streri, 2000). Later sensorimotor developments in sitting up, standing and moving about freely allow fast quasi-automatic visual orienting to other body locations that are being touched.

More recent work with newborns also suggests that inputs from different modalities interact to some extent right from the start. Human newborns recognise some aspects of shape across vision and touch. Streri and Gentaz (2003) found that newborns look longer at a new shape that they had not felt in their hand before than at a shape that had previously been placed repeatedly into their hand. Some intersensory effects thus seem to be present at birth in humans. Indeed, there are many more neural connections across different areas of the infant brain early on than later. Many neural connections drop out with lack of stimulation and greater differentiation with experience (Huttenlocher, 2002; Katz & Shatz, 1996). There is also evidence that multisensory systems – for instance, in the midbrain superior colliculus – only develop gradually in the postnatal period in primates (Meredith & Stein, 1996; Stein, Wallace, & Stanford, 2000; Wallace & Stein, 2007). Both decreases and increases in crossmodal connections with development must be assumed.

Evidence for brain plasticity in neuronal connections (e.g. Katz & Shatz, 1996) was an important consideration also in considering early visual deprivation (see later).

Verbal mediation (Ettlinger, 1967) is clearly not a necessary pre-requisite for crossmodal effects, as had sometimes been assumed (Ettlinger, 1967), although crossmodal effects can, of course, also occur through learning. Seeing the name "green" written in red causes a well-known increase in recognition time.

There was also evidence against the notion of a learned non-verbal translation system. Neither age nor stimulus complexity affected crossmodal matching more than intramodal matching, as would be expected if learned translation had been involved. The relation of crossmodal to intramodal matching was inconsistent in different studies. In some studies, particularly with shape stimuli, crossmodal matching was at least as good as and often better than intra-tactual matching (Cashdan, 1968; Millar, 1971, 1972a; Milner & Bryant, 1970; Rudel & Teuber, 1964). In others, especially with length stimuli, crossmodal performance was worse and less consistent than performance within the modalities (Connolly & Jones, 1970; Jones & Connolly, 1970; Kress & Cross, 1969, Millar, 1972b). The discrepancies could not be explained simply by the difficulty or complexity of the stimuli. More complex or double distances were more difficult, and younger children produced more errors. But neither complexity nor age interacted with the modality conditions (Millar, 1975a). Apparently trivial effects of the order in which visual or haptic standards were presented were responsible. Moreover, asymmetric results in the opposite direction had been found in similar studies previously (Connolly & Jones, 1970; Jones & Connolly, 1970). Some modality-specific differences in immediate memory for the two modalities seemed to account for these results.

The findings were thus broadly consistent with the view of both Eleanor and J. J. Gibson (see Chapter 1). But the fact that the same two crossmodal inputs produced significantly asymmetric effects could not be explained solely in terms of detecting the same information from visual and kinaesthetic cues. Nor could it be assumed that modality-specific factors are irrelevant.

Short-term memory for unfamiliar serial stimuli tends to be much briefer and/or less accurate for serial tactual stimuli than in vision (Goodnow, 1971; Millar, 1972a, 1972b). Memory spans for tactual dot patterns that young children could not name quickly were at most for series of two or three items only. Children achieved higher spans corresponding to the speed of their naming the patterns in pre-tests (Millar, 1975b, 1975c). Fast recoding into a form that allows articulatory/phonological rehearsal across short delays increases memory spans (Baddeley, 1986, 1990). Unfamiliar visual shapes were also more easily imagined across delays than unfamiliar felt shapes by sighted preschool children. Preschoolers' memory for unfamiliar shapes increased across short delays when they were asked to try and "see the shape their head" during the brief delay (Millar, 1972c). Recoding into a more parsimonious form (Miller, 1956), albeit not one that necessarily involves semantic reorganisation, is probably involved. The "visuo-spatial sketchpad" that is assumed in Baddeley's working memory model as a further adjunct to a central executive control process is assumed to involve some covert activity (Goldman-Rakic & Friedman, 1991; Logie, 1995; Smyth & Pendleton, 1990). Moreover, later versions of the model also propose a further system that assumes that spatial recoding does not apply to vision alone (Baddeley, 2000, 2007).

We have since found that children's recall spans for a series of different blind movements tend to be relatively small (Ballesteros, Bardisa, Millar, & Reales, 2005). However, immediate haptic memory for the length of a hand movement has to be distinguished from a series of differently shaped hand movements, and from recall spans for a series of unfamiliar tactual patterns (Millar, 1975b). Moreover, movement information can be kept in mind during short delays by being "rehearsed" covertly, even by young children in blind conditions (Millar & Ittyerah, 1992).

Mental rehearsal of blind movements thus resembles the articulatory rehearsal that has been demonstrated in working memory for verbal or phonologically coded materials (Baddeley, 1986, 1990). However, instructions to rehearse blind movements covertly may be less effective than articulatory coding, at least for young children, unless the start and stopping points of the criterion movement can be coded spatially (Millar, 1999a, 2000).

Concomitant bimodal inputs from vision and touch benefit purely haptic matching (Millar, 1971). Inputs from vision and touch that overlapped in presentation produced better haptic matching than when the information came solely from touch.

The importance of concomitant bimodal inputs was shown in the seminal study on visual and auditory (speech) effects by McGurk and McDonald (1976). People heard an amalgam of the syllable that was fed into their ears and the different syllable that was being formed by the lips of a face they saw. The advantage of lip-reading when hearing is impaired is now well known (Dodd & Campbell, 1987; MacDonald & McGurk, 1978).

Evidence for convergence and partial overlap of multimodal inputs has been growing rapidly in the last fifteen years. Most studies have been on vision in conjunction with auditory and speech effects, and the convergence of vision and hearing in spatial tasks. But a growing number of studies now include touch as well as vision (Calvert, Spence, & Stein, 2004). Effects of passive touch have been shown, especially with respect to orienting or attentional responses to stimuli in another modality (Spence & Driver, 2004).

In the spatial tasks considered here the focus is on the relation of vision to active touch, or haptic perception, which includes exploratory movements, rather than on vibrotactile stimulation or passive touch. Vision tends to be more accurate and faster than haptic perception. It often "dominates" when haptic cues conflict with vision in shape and spatial tasks. Task demands and conditions can reverse that (van Beers, Wolpert, & Haggard, 2002; Ernst, Banks, & Buelthoff, 2000). Vision can be said to be more specialised than touch for spatial information in humans and other primates. But that does not mean that information from vision is always more effective than information from touch and movement (van Beers, Wolpert, & Haggard, 2002).

Touch, with its specialised receptors for object properties such as temperature, resistance and elasticity, seems to be as good as vision for identifying objects (Klatzky, Lederman, & Reed, 1989). It is also as good as – or can indeed be better than – vision at detecting finer surfaces textures (Heller, 1989, but see Guest & Spence, 2003). The point is that shape features and spatial information are also perceived by touch, especially in conjunction with movement information (van Beers, Wolpert, & Haggard, 2002).

It is reasonable to assume that diverse, specialised sense organs evolved in higher organisms because they benefit from the additional information these provide. Some overlap and convergence of the diverse inputs would be necessary, if only to ensure consistent perception and action (Pochon et al., 2001).

But receiving converging or overlapping inputs about the same goal or target from more than one source also has additional benefits The apparent "redundancy" of information from converging and overlapping cues enhances the available information, particularly in conditions of uncertainty. Such redundancy can also act as a "fail-safe" device in case of injury to one of the diverse sources or brain regions that are most involved in the performance of particular tasks.

NEUROPSYCHOLOGICAL EVIDENCE AND CROSSMODAL EFFECTS

If anything, neurophysiological studies show even more clearly that inputs from diverse sources converge and partially overlap. It is worth reminding ourselves first, if only briefly and sketchily, of the main specialisation of the primate brain into regions and sub-regions that primarily serve inputs from a given sensory source, but also have many connections with other regions.

The occipital lobes contain the visual areas. Light stimulation from the rod and cone receptors in the retina of the eye is relayed via the optic nerve. The nerve fibres divide at the level of the optic chiasm and proceed via the lateral geniculate nucleus to primary and secondary striate areas. Cutaneous and somaesthetic impulses reach somatosensory areas in the anterior parietal cortex via two pathways: The spinothalamic pathways mainly carry impulses from pain, temperature and diffuse touch receptors to the reticular formation, thalamus and cortex. The lemniscal system transmits impulses from tactile and kinaesthetic sources via the spinal cord and dorsal columns across synapses

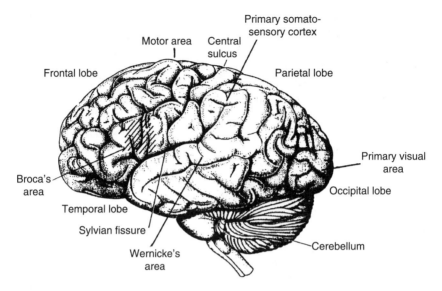

Figure 2.1 The left hemisphere of the human brain. [Adapted from S. Zeki (1993), *A Vision of the Brain*. Blackwell Scientific Publications, Oxford. Reproduced with permission from the author and Wiley-Blackwell Publishing Ltd.]

in the medulla and a thalamic relay nucleus. The motor area is concerned with finger, hand and limb movements and is highly organised in a head-to-foot direction. The area directly below it is concerned with articulation. Areas in the temporal lobe of the left cerebral hemisphere are specialised for hearing and language skills and are activated more than the right hemisphere in language tasks (Figure 2.1). For spatial tasks the right hemisphere tends to be more specialised, particularly, though not solely in the parietal areas. These also connect to the prefrontal cortex, which is particularly important in humans.

Figure 2.2 shows the right cerebral cortex from the inside, without the left cerebral cortex. The two cerebral hemispheres are connected by a network of fibres in the corpus callosum. The primary areas serving inputs from different sensory modalities are organised in depth, connecting peripheral sense organs to the cortex via subcortical regions, but also laterally, connecting different regions of the cerebral cortex.

Spatial perception and performance depends on a number of distributed cortical and subcortical areas, which have intricate and mostly reciprocal interconnections. It is noteworthy that these areas receive inputs from multiple sensory sources (Andersen, 1999; Andersen, Snyder, Bradley, & Xing, 1997; Duhamel, Colby, & Goldberg, 1991; Gurfinkel & Levick, 1991; Sakata & Iwamura, 1978; Stein, 1991; Stein & Meredith, 1993). The evidence has come from single-unit recordings from neurons in the monkey brain, from

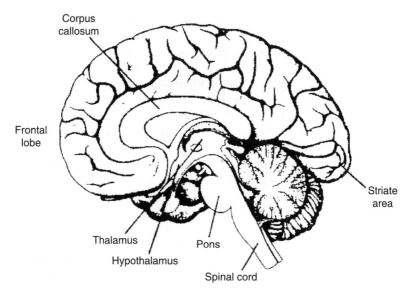

Figure 2.2 The right hemisphere of the human brain, seen from the inside (without the left hemisphere). [From S. Zeki (1993). *A Vision of the Brain*. Blackwell Scientific Publications, Oxford. Reproduced with permission from the author and Wiley-Blackwell Publishing Ltd.]

clinical cases of people with cerebral damage who are impaired on spatial tasks, and increasingly now also from studies that explore the human brain with non-invasive techniques.

Single-unit (neuron) recoding from the monkey brain showed, for instance, that discriminating straight and round objects held in the hand involved different converging inputs from the skin and from proprioceptive inputs from finger joints, which combined in single neurons in the parietal cortex (Sakata & Iwamura, 1978). The specificity of neural cells that respond only to particular forms of stimulation is not complete. Some cells in brain areas concerned with spatial processes respond to both visual and tactual stimulation and are likely to be involved in integrating external spatial cues (Graziano & Gross, 1994). Bimodal neurons may also be important in the "personal space" that immediately surrounds the body (Graziano & Gross, 1993).

Visual, proprioceptive, vestibular, limb movement, as well as auditory and motivational signals come together in regions of the posterior parietal cortex that are involved in spatial processing. Circuits between the superior parietal lobe and the dorsal premotor cortex are critical for reaching and for locating the hand (Andersen & Zipser, 1988) and in spatial processes that are involved in the control movements (Berthoz, 1991; Gurfinkel & Levick, 1991; Paillard, 1991; Pochon et al., 2001). Another network has been suggested in transferring adaptive control in coordinating posture and movement from cerebral to cerebellar subsystems (Burnod & Dufosse, 1991). The hippocampus is involved in associative and spatial short-term memory and has reciprocal connections with many areas of the neocortex (Bayliss & Moore, 1994; O'Keefe, 1991; O'Keefe & Nadel, 1978; Rolls, 1991). The prefrontal cortex, which is important in human cognitive, inferential and problem-solving tasks, is also implicated in spatial processing. It has connecting circuits with the parietal cortex and hippocampal regions and also has reciprocal connection with the subcortical areas that are involved in arousal and motivational states (Mesulam, 1998).

The interrelations between cortical and subcortical brain regions that are concerned in processing spatial information in different types of tasks were described in terms of networks of neurons (Arbib, 1991; Fessard, 1961; Stein, 1991, 1992; Stein & Meredith, 1993). Description in terms of a network of interrelating factors seemed to me appropriate also for describing behavioural findings on crossmodal and spatial behaviour. Indeed there is now increasing evidence for describing brain functions in terms of networks of neurons that interrelate (e.g. Gazzaniga, 1994; Levine, 2002; Rizzolatti, Fogassi, & Gallese, 1997, 2004).

All the brain areas mentioned earlier are subdivided further. The posterior parietal cortex seems to exemplify the general principle that high levels of spatial performance involve both more specialisation in the circuitry and also wider connections with other areas. Populations of neurons nearer to visual areas respond in visuo-spatial tasks. Populations of neurons nearer to somatosensory areas respond to haptic spatial tasks. Reaching a target from a changed starting position involves different regions in the parietal cortex of the monkey from those involved in non-visual reaching from a changed starting position to a target that is defined by reference to arm positions (Rushworth, Nixon, & Passingham, 1997a, 1997b).

The dual function of specialisation together with overlap and apparent redundancy thus seems to be a characteristic rule of the organisations of the central nervous system and brain of primates.

The old dichotomy that the senses are either separate and provide quite different information, or else are unitary and all provide the same information, could thus be laid to rest. Taken together, the findings seemed best explained in terms of the convergence and partial overlap of inputs from diverse sources.

However, the very advantages of an overlap of visual and haptic information raised questions about haptic perception without vision.

VISION AND BLINDNESS

It is not easy to counter the notion that vision underlies haptic shape and spatial perception, or that vision plays an integral role. Evidence from studies with blind children and adults has been inconsistent, even paradoxical. There are a number of reasons why the findings are not always easy to interpret.

An important factor in the apparent inconsistency is the question that is being asked. The question whether blindness eliminates a person's spatial "ability" – that is to say, whether blindness eliminates the person's potential for acquiring and using spatial information – differs from asking how total absence of information from vision affects performance on spatial tasks (Millar, 1994). The distinction is important, but it is often forgotten in interpreting empirical findings. There are also a number of methodological constraints. Total lack of vision from birth, uncomplicated by other impairments, is actually very rare indeed in normally able blind populations, at least in "developed" countries.

Helen Keller is probably the most famous example of general competence in total blindness. It is less generally known that she was sighted until the age of 18 months. A blind man studied by Gregory and Wallace (1963) was able to recognise shapes by vision without explicit training after his sight was restored by cataract operations in middle age, although he was insensitive to some visual cues. The man may have had some sight until the age of about ten months, when he was diagnosed as blind. It is not entirely clear whether he had any residual vision before cataract removal that may have helped in riding a bicycle before the operation. But his haptic experience with uppercase letters and telling the time from a tactile watch also enabled him to recognise similar visual inputs.

Length of visual experience prior to blinding was an important variable in pioneering studies on blind children and adults. In many instances, "early blindness" meant blindness before the age of about three years. It was compared with "late" blindness that meant either becoming blind after childhood or recent loss of sight. But it shows that even "early blind" participants were often by no means visually naïve. Very young infants can discriminate shapes visually, and the human visual system develops relatively fast (Maurer, Lewis, Brent, & Levin, 1999). By the age of six months a sighted baby has considerable visual experience not only in looking at and perceiving shapes, but also in reaching for, batting and trying to grasp objects and in attempting to sit upright in the visual world.

Many totally blind people have thus had visual experience in the early years. Results showing that shape perception and spatial performance was nevertheless often worse for the early blind than for people with longer visual experience seemed to imply that vision is needed to integrate spatial information and underlies spatial performance (e.g. Warren, 1977; Worchel, 1951). The number of young blind people is mercifully very small in Western Europe. Studies comparing differences in visual experience therefore necessarily needed rather broad banding in each category of early versus late experience for there to be sufficient data to make statistically valid comparisons.

However, it has to be recognised that differences in levels of proficiency between quite small groups may simply be due to differences in the amount and type of spatial information that has been made available to the blind in the past. It cannot be taken as evidence that the early blind have less spatial "capacity" or potential for acquiring spatial information than do late blind people (Millar, 1988a).

A similar problem has to be kept in mind for results that depend on comparing blind people with blindfolded sighted cohorts. Matching cohorts to a very small group of blind people on standardised tests is not always easy, if only because standardised tests for the two groups will necessarily have been standardised on different representative samples. That can be overcome to some extent by using items that the tests have in common, such as digit recall and reversed digit recall. Even so, lower spatial scores by a small group of blind people than by their sighted cohorts is not, as such, evidence that they differ in their potential for acquiring and using spatial information when the relevant spatial information is made available to them.

By no means all studies have shown worse performance by blind adults than by their sighted cohorts in tasks that require spatial inferences (e.g. Carpenter & Eisenberg, 1978; Hollins, 1986, 1989; Klatzky, 1999; Loomis et al., 1993; Marmor & Zaback, 1976). It is not always clear that the necessarily small groups of blind participants consist only of blind people who had never had any visual stimulation at all.

However, in principle, even one positive result is sufficient to show that the blind do not lack the "ability" or potential to do spatial tasks (Evans, 1985). But that stricture applies only to questions about spatial "ability" or the potential for acquiring spatial information. It is not relevant to questions about effects of the available information on performance (Liben, 1988).

Some quite recent studies are still concerned to show that visual experience is not necessary for solving spatial problems that involve understanding spatial layouts (e.g. Tinti, Adenzato, Tamietto, & Cornoldi, 2006). Strong empirical evidence for separating visual and spatial coding also in vision was reported by Klauer and Zhao (2004). They showed that visual short-term memory was disrupted more by visual than by spatial interference tasks during delays, whereas spatial short-term memory tasks were disrupted by spatial rather than by visual interference.

The important point made in the study by Tinti and colleagues, and in the recent argument by Cornoldi and Vecchi (2003), is that covert strategies or forms of mental representation are not necessarily identical for all blind participants. Indeed, that may be said also about the sighted. Forms of imagery that are derived from other modalities are reported also in blindness (Cornoldi & Vecchi, 2003; Schlaegel, 1953).

The very possibility that the sighted are able to use visual imagery is sometimes adduced as a further reason for assuming that visual experience is essential in spatial tasks. Many, but by no means all, sighted people report using visual imagery when they have to perform haptic tasks blindfold. But findings vary with task conditions. More sophisticated methods support reports from people who say that they visualise haptic stimuli (e.g. van Beers, Sittig, & Gon, 1999; Kosslyn, 1987, 1994; Lederman, Klatzky, Chataway, & Summers, 1990). Moreover, a neural basis for visual imagery was suggested by studies with non-invasive techniques (Goldenberg, Podreka, & Steiner, 1990).

These findings do not, of course, address the historical "imageless thought" controversy that has lasted into our time. But it is debatable whether it is necessary to assume that thinking must be described either in terms of imagery or in terms of abstract propositions. People's covert heuristics are not necessarily the same for all types of problems or for all forms of information.

Differences in the type and amount of spatial information that is available to an individual, whether blind or sighted, seemed to me to explain the discrepancies in empirical findings most simply, in line with the suggestion by Leonard and Newman (1967) that whatever difficulties congenitally blind people may experience, problems of spatial orientation are more likely to be caused by lack of experience than by blindness as such.

At the same time, the relatively poor performance in spatial tasks with inputs from touch and movement compared to vision, and the advantages for touch of getting "redundant" information from vision, raised the question of how haptic processing of shape and spatial information takes place in the total absence of sight.

EXPERIMENTS WITH BLIND CHILDREN

I undertook research with blind children to try to answer the question of how haptic inputs are coded spatially in the total absence of sight, but also because the findings would have significant practical implications.

At that time, blind children were taught in special schools, most of whose teachers had been specially trained to teach blind children. I was allowed access to school records for the research, which enabled me to ascertain reliably the potentially significant factors of years in blind education, age, and the type, degree and length of blindness.

One thing became immediately obvious. Only a minute proportion of the blind children had been totally blind from birth. Consistent with previous findings, young congenitally totally blind children were indeed often significantly poorer on unfamiliar spatial tasks than blindfolded sighted children, or children who had been blinded later.

The differences between blind and sighted children that were found depended on the task, but even more on the information that was available to them. Congenitally blind children performed as well as blindfolded sighted cohorts on "yes/no" haptic recognition of small nonsense shapes. They were faster, although somewhat less accurate (Millar, 1974). They did as well as the sighted, and as children who had been blinded later on, in recognising the orientation of a vertical line that was aligned to the child's body midline in tabletop space or in large-scale space (Millar, 1976).

The important difference was thus not so much that congenitally totally blind children performed less well than the later blind or the sighted, but that their performance patterns tended to differ. Their errors occurred predominantly in conditions that disrupted the relation between the location of an object felt previously and their body posture.

Such tasks involved a change in the location of a stimulus object relative to either an external surround or to the observer's upright posture, or a change in the observer's position or posture relative to a stimulus array. Other tasks that disrupt coding involve changes in the alignment of stimuli relative to the observer's body midline, or changes in the direction of movements that are needed to access a target location (Millar, 1975d, 1976). The difference showed even in a simple task of moving small toys from the corners of a square- or a diamond-shaped plate to the corners of an identical square or diamond. Blindfolded sighted children made more errors with diamond- than square-shaped backgrounds, regardless of whether the test shape was located on the other side of the child's body midline, or above the relevant square- or diamond-shaped plate in tabletop space. They coded the toy locations by reference to the external shape of the plate. Errors by congenitally totally blind children did not differ between the two types of shape. Far the largest proportion of their errors occurred when the standard and test plates were placed on either side of their body midline, so that their hand movements in placing the toys at the corresponding location crossed the body midline. They were significantly more accurate when vertical placement movements were needed that preserved the relation of the corner locations to the body midline (Millar, 1979).

Blind children produced a very high proportion of location errors when a previously body-aligned square display was inverted. The errors consisted predominantly of indicating the position of the target location on the original, now incorrect, side of the display, suggesting that they had coded the target location by reference to their own position, although the rotation task had been carefully demonstrated. Blindfolded sighted children were also less accurate under rotation. However, unlike the congenitally totally blind, their errors consisted mainly of locating the target inaccurately on the correct side of the inverted display. The errors by the blindfolded sighted suggested that their coding heuristics made it easier to update the location of the target, albeit not very accurately (Millar, 1981a).

No one doubts that experience and practice are important. But that commonplace is often forgotten when it comes to interpreting differences shown by "special" populations. Repeated information from scanning movements made a significant difference for both blind and blindfolded sighted children even with displays that had not been rotated. Changing the origin of positioning movements for locating the target produced significant location errors (Millar, 1981a, 1981b). Changing a child's posture to a sideways position increased the movement bias very significantly for congenitally totally blind children, and more so for the younger blind than for the older blind. Young blindfolded sighted cohorts showed the same very significant movement bias in recalling the end location of the movement. But their errors were not significantly increased by the sideways posture. The movement biases virtually disappeared when blindfolded sighted children were instructed to feel the square surround of the display and to remember the critical location in relation to that external square surround (Millar, 1985a).

Rotation tasks are more difficult than simple recall tasks for sighted observers as well as for blind observers. The task conditions require people to mentally update the location of a stimulus object. They thus impose a greater cognitive or memory load. In principle, using Euclidean geometric principles should achieve accurate performance, regardless of modality. In practice, adults and children are influenced by the information provided by their habitual mode of spatial experience (Rieser & Heiman, 1982; Rieser, Lockman, & Pick, 1980).

Heuristics suggesting that people attempt to visualise a spatial array and try to imagine the rotation of the object in an external context are implied by the length of time it takes people to arrive at correct decisions (Kosslyn, 1987; Shepard & Metzler, 1971). But visuo-spatial strategies in mental rotation tasks are by no means necessarily accurate. Seeing the original position of the object relative to the external surround can produce typical visuo-spatial errors, suggesting that viewing the actual position interfered with trying to rotate it mentally.

Moreover, some apparently purely visual effects, such as the "oblique" effect, which names the finding that vertical and horizontal directions are judged more accurately than oblique directions, are also found in haptic conditions. The oblique effect varies with gravitational cues, body-and-head position and tilt, suggesting that combinations of different reference cues in touch can produce the same or similar effects (Gentaz & Hatwell, 1996, 1998).

However, visual conditions do have the advantage that external and body-centred reference cues normally coincide and overlap in our gravitationally oriented environment. In the total absence of sight, concomitant external background cues relative to a felt object are rarely present, unless the potential external cues are already known and can be sought. External sounds can be used to indicate locations, especially when such strategies are reinforced. Knowing what the sounds mean makes them useful as orienting or reference cues for updating movements in familiar environment. Mobility teachers instruct blind children in their use. But unfamiliar sounds do not provide reliable external guides to locations in new environments in blind conditions. The total absence of inputs from one source thus tends to produce an imbalance in the type of spatial information that is available from different sources.

The lack of redundancy in the type of spatial information that is normally available in the complete absence of sight is thus potentially a major drawback that is not always recognised sufficiently. In familiar contexts, recognising a sound allows the blind child to adjust his or her heading movements towards or away from a known goal, or to move towards a corner of the path that he or she can feel underfoot as a landmark that signals a turn. That is not the case in new environments.

What is missing in the total absence of sight is thus not merely substitute information from another modality. The habitual overlap of external and body-centred spatial cues that concomitant vision normally provides also needs to be reconstituted in the total absence of sight from birth. Young children need more informational redundancy in any case (Gollin, 1960; Millar, 1986, 1994; Munsinger, 1967). Lack of vision increases uncertainties. The need for informational redundancy from other sources is therefore greater. The practical implication for rehabilitation is that substitute information alone is not enough. It is equally important to restore the overlap between different input sources that is depleted in the total absence of a major sensory source (Millar, 1994).

The main effect of total absence of sight from birth on haptic spatial performance was a greater reliance on body-centred and movement cues as a means of remembering the locations of objects and directions, suggesting a relative lack of redundant converging information from external reference cues. Such findings have sometimes been treated as if they were statements about the spatial "ability" of blind children. To do so is to confuse experimental findings with interpretations (Millar, 1999a). Thus, the results of a recent study are completely consistent with previous findings which suggest that early blindness can decrease people's reliance on external reference information (Röder, Kusmierek, Spence, & Schicke, 2007). Even the availability of spatial vision in front of us, compared to its lack behind our backs, makes a difference. Spatio-temporal order judgements for passive tactual stimuli, delivered to the fingers of crossed hands, are less disturbed if the hands are crossed behind the back than in front (Kobor, Furedi, Kovacs, & Spence, 2006).

The findings actually make very good sense in terms of the spatial information that is actually normally available in the total absence of sight. Body-centred reference cues are not only present, but they are also much more reliable than information about external cues in blind conditions. Relying on body-centred coding can make it more difficult to mentally update one's own position, or the relative position of objects in typical mental rotation tasks. However, there is no reason to doubt that it is also perfectly possible in principle to apply geometric rules to body-centred reference axes, or to use external reference cues when that information is available in addition or with experience (Hollins, 1986).

The apparent paradox that congenitally totally blind children are often severely impaired in spatial tasks, but that absence of vision is no bar to spatial performance that parallels, and may even exceed, that of the sighted is resolved when differences and/or alternatives in the available relevant information are taken into account.

The results and the interpretation were more consistent with the newer neurophysiological evidence on the much greater plasticity of brain organisation than with previous models, which implied a more rigid architecture.

CEREBRAL PLASTICITY AND BLINDNESS

I mentioned von Senden's (1932) totally negative results on people's visual perception after an operation for dense cataracts dating from birth that was expected to allow them to see. The results may have been caused partly by the techniques then available for removing cataracts. But later studies did indeed show that the visual system requires stimulation and patterned visual inputs to develop normally (Chow, Riesen, & Newell, 1957; Maurer et al., 1999; Riesen, 1947; Shatz, 1992; Wanet-Defalque et al., 1988).

Total deprivation of sight from birth produces molecular as well as molar changes. Metabolic activity and the uptake of glucose are higher in the visual cortex of early blind compared to later blind people (Veraart et al., 1990; Wanet-Defalque et al., 1988). But these may also be concerned in producing compensatory changes.

Kujala and colleagues showed enhanced processing of sound location in the early blind (Kujala, Alho, Paavilainen, Summala, & Naatanen, 1992), and they later also showed visual cortex activation in blind people during sound discrimination (Kujala et al., 1995). Improved auditory spatial tuning in blind humans has been demonstrated (Röder et al., 1999). Studies using event-related potential (ERP) techniques found similar activation in the occipital (visual) cortex during tactile as well as auditory tasks (Rösler, Röder, Heil, & Henninghausen, 1993).

It has been suggested that compensatory brain reorganisation may come about either through increased stimulation of other sensory areas, including somatosensory areas, or by changes in neural connections (Hyvärinen, Carlson, & Hyvärinen, 1981; Hyvärinen & Poranen, 1978; Leinonen, Hyvärinen, Nyman, & Linnankoski, 1979; Morris, 1989). These connections also concern the parietal association areas that are involved in spatial processing. Evidence that increased haptic experience also has compensatory effects that affect visual regions of the cortex raises questions that will be considered later, particularly with respect to braille.

The neurological evidence on cerebral plasticity and the importance of converging and partly overlapping inputs from different sources even at the time correlated better with my findings on spatial coding in the total absence of vision than previous more rigid models of the brain and brain functions.

But the finding that congenitally blind children tended to rely more on body-centred reference cues also highlighted the importance of different types of reference cues in explaining spatial performance.

THEORETICAL ASSUMPTIONS: INTEGRATIVE PROCESSING OF INPUTS FROM DIVERSE SOURCES

The aim of the theoretical description of how spatial coding takes place was to account for the varying of effects of crossmodal and bimodal inputs, and for the apparent discrepancies in effects on haptic spatial perception and performance in the total absence of sight from birth, suggested by behavioural and neurophysiological studies.

The description assumes that sensory inputs provide specialised, but complementary and convergent, information. That includes the important systems that signal gravitational directions. It was argued that the form of relational organisation that spatial processing produces varies with the overlap and reliability of converging inputs from diverse sources that are available for integration in different task conditions (Millar, 1994).

The description was summarised as "convergent active processing in interrelated networks" (CAPIN, for short). The terms indicate the main characteristics suggested by the behavioural as well as by the neurophysiological evidence.

Convergent processing labels the process of integrating converging sensory inputs from diverse systems into the spatial organisation that spatial tasks require. As noted earlier, the notion of interrelated networks provided a better description of the complex interactions between cortical and subcortical areas in spatial processing than the previous assumption of a rigid peripheral-to-central direction of processing. The description also accounts more easily for the apparent discrepancies in spatial coding with differences in information at the behavioural level.

Spatial coding was described as an active process. It need not be assumed that people are aware of all, or indeed of any, of the processes that contribute to perceiving objects in relation to each other, or to a particular scene, or in ascertaining the orientation of a felt object by reference to body-centred cues. Passive touch can elicit automatic orienting responses to the relevant location on the body without our being aware of the processes that made this possible.

Haptic perception involves active exploratory movements. But the findings also showed that shape and spatial processing of inputs from touch and movement in the absence of external cues from vision depends more on processes that relate the inputs to body-centred cues.

The finding highlighted the fact that it is actually impossible to perform spatial tasks without reference cues that specify the location, distance and direction of objects relative to each other, or to an external surround, or to body-centred cues in large-scale as well as in small-scale space.

The proposed model thus assumed that the crucial aspect of spatial processes is that they integrate inputs from all available sources, in terms of reference cues that can specify the location, distance or direction of objects that we perceive or need to reach or to find. The form and efficiency of the reference organisation depends on the redundancy or overlap of compatible cues that are available for integration as possible reference cues (Millar, 1994).

But it also raised the question how far the implications about the specifically spatial aspects of spatial tasks and of spatial information can be used operationally in a form that can apply generally in experimental studies.

WHAT IS SPECIFICALLY SPATIAL ABOUT SPATIAL INFORMATION?

Curiously enough, there are no agreed explicit criteria for what is to count as the specifically spatial aspect of a task or of spatial information. It is usually assumed that we know what is meant by a spatial task, so that explicit criteria are unnecessary. The frequent identification of vision with spatial perception may have contributed to that. It is less obvious for spatial perception through touch and movement.

The nearest to an agreed criterion is the distinction between locating "where" an object is and identifying "what" it is (Ungeleider & Mishkin, 1982). The distinction is based on evidence for neural circuitry that describes a "dorsal path" involving posterior parietal and other areas of the brain that are important for spatial activities, whereas "ventral path" neural streams are involved in the identification of objects. The distinction is considered further in a later chapter.

Location, or "where", tasks are useful for testing the efficiency of spatial performance. But they do not answer the question of how visual or haptic spatial performance is achieved. For that we need to specify what aspects of inputs from any modality are crucial for perceiving and finding the location of objects and, indeed, for perceiving, recognising and remembering distances and directions, which spatial tasks also involve.

In fact, however, the assumption that spatial processing crucially involves the integration of inputs, from whatever source, in some form of reference organisation is implied in almost all spatial theories.

Descriptions of spatial performance are frequently couched in terms of "frames of reference" (Berthoz, 1991; Danziger, Kingstone, & Wards, 2001; Holdstock et al., 2000; Keulen, Adam, Fischer, Kuipers, & Jolles, 2002; Howard & Templeton, 1966; Paillard, 1991), "cognitive maps" (Tolman, 1948), different types of "spaces" (e.g. personal space, external space), or coordinate axes with different bases. Even theories which assume that vision underlies spatial integration usually imply that this is because vision provides external or environmental frames.

The problem is that terms such as "frames of reference" and "cognitive maps", together with suggestions of "unconscious inference" (Helmholtz, 1867, 1896), if taken too literally as "cognitive structures" of the type that aroused Gibson's ire (Chapter 1), sometimes enter into controversies over "direct" versus "indirect" perception. The problem disappears if the terms are considered shorthand for describing the relational reference organisation of inputs from diverse sources as an activity of the organism.

Berthoz (1991) stressed the importance of multiple reference frames for spatial orientation and the perception and control of movement. He considered the term to be equivalent to "a set of values to which each neuronal variable can be referred" (p. 81) and assumed that they depend on interactions between inputs at many levels of an organism's neural system.

Spatial processing is generally classified into three very broad categories by their main sources of origin. Body-centred or "egocentric" reference cues designate cues arising from gravitational, posture and proprioceptive cues within the body. External, "extrinsic" or "allocentric" reference cues designate cues that originate in the environment or surrounds of targets. Object or shape-based reference cues also arise from stimuli that impinge on the organisms from external sources. The spatial relations here are between features within shapes relative to each other. A salient feature within a shape can act as a reference cue relative to other features or to the total configuration. As far as I know, there is no disagreement about this very broad classification of different forms of spatial relations.

I consequently defined spatial processes as processes that integrate and organise the available inputs from diverse sources to serve as reference cues that specify the location ("where"), distance ("how far") or direction ("what turns") for perception and action that a given spatial task demands (Millar, 1994).

The definition assumes that spatial tasks require the integration of available inputs in terms of reference organisations. In principle, it covers diverse inputs from all available sources that can serve as potential reference cues.

Although it has not been used previously as a defining criterion, experimenters can and do vary inputs from the different sources that can potentially serve as reference cues to test their effect on performance in a given spatial task (e.g. Rieser, 1979). It could thus serve as a general operational definition.

QUESTIONS RAISED BY THE PROPOSED MODEL

The criterion, henceforth called the "reference hypothesis", was tested further in the studies that are considered in the context of the questions discussed in the subsequent chapters.

The first question concerns effects of extraneous and body-centred cues on walking through large-scale space that is devoid of reference information. The second concern, by contrast, is about the role of finger and hand movements in the pick-up of information from the small raised dot patterns that constitute the braille cell and from somewhat larger raised spatial layouts in table-top space. Findings are reviewed which show how the physical composition and small size of braille patterns make it difficult to use shape-centred or body-centred cues to code the patterns as outline shapes, and how task requirements and experience affect the movements and functions of the two hands relative to each other and to the text. The function of scanning movements by the left and right hands in the pick-up and coding of spatial information from larger raised layouts is considered next also in the context of explicit instructions to use body-centred and external cues.

The further prediction that effects of external reference cues can also be demonstrated in purely haptic conditions is considered next. The question of what, if anything, touch and movement have in common with vision is particularly intriguing for theories of space perception and performance. It is discussed in the context of three different shape illusions that occur in touch as well as in vision. The frequent assumption is that the similarity is purely fortuitous. The predictions here are that similarities are due to similar discrepancies in the reference information that normally produces accurate performance. Experiments that test the predictions are detailed in the context of previous evidence.

The final predictions concern the question of whether added vision improves haptic performance regardless of whether it provides relevant spatial reference cues. Findings from experiments that test predictions from the reference hypothesis are discussed in the context of the view that vision facilitates haptic performance even when it is irrelevant.

Evidence that relates to the different questions is discussed in turn in the chapters that follow.

CHAPTER THREE

Reference cues in large-scale space

This chapter considers effects of reference cues in large-scale space. The reference hypothesis assumes that the direction of walking through large-scale space depends crucially on the inputs that are available from diverse sources for integration as reference cues that update the direction of locomotion.

Large-scale spaces differ more in the tasks that they set the traveller than is often realised. The chapter examines first the kind of information that different types of tasks and spaces afford and how the traveller's resources contribute to their solution in the presence and absence of sight.

An important prediction from the reference hypothesis is that the direction of locomotion in large-scale spaces that are totally devoid of reference cues depends on using initial posture cues in conjunction with whatever extraneous cues occur and/or with changes in body posture. Findings from studies in such conditions are reviewed, before considering the results of a study that was specifically designed to test the hypothesis. The study used a space from which all reference information was carefully excluded, so that effects of extraneous cues and of body-centred cues could be manipulated experimentally. The implications of the findings are then considered further.

INITIAL SIGHT OR SOUND OF A TARGET OR GOAL

Perhaps the simplest task for normally sighted adults in large-scale space is to walk straight to an object or target that they see in the environment. Young children, once they can toddle, do not need special lessons to walk straight to where they want to go. Even built-up spaces can provide a distant view of the goal to be reached. You can see the Eiffel Tower from quite a long way away along the Champs Elysées.

Sighted adults and children can also reach new locations blindfolded if they have had prior sight of the environment (Farrell & Thomson, 1998; Rieser, 1979; Rieser, Ashmead, & Youngquist, 1990; Rieser, Guth, & Hill, 1986). Accuracy in reaching the goal declines with the length of the delay interval between seeing the target and moving to it in the dark (Rieser & Rider, 1991).

A sound from a known target can function similarly as initial sight of the target. People can walk straight to the source of a heard command for a short distance without sight if their body posture is oriented to face the sound at the start, and so long as that posture is maintained.

That, at any rate, was the finding with blindfolded sighted and congenitally totally blind children in the control condition of the experiment that will be detailed later (Millar, 1999b). The experiment took place in a quiet room that was unfamiliar to the children. The children were told to walk in a straight line from their starting point to the experimenter, whose voice they knew well, at the command "Come". Extraneous sounds were controlled. The floor was carpeted, so that they could not hear their footfall, and the walk started from several different directions in the room. The important factor was to get the child to stand straight and to ensure that the body, head and leg postures all faced the potential target location. All the children walked in a straight line to the experimenter in that condition. They made no mistakes, even when they had been told to wait several seconds after the voice had stopped before starting to walk.

The crucial factor thus seems to be to maintain the body posture that was assumed in facing the target sound. That enables the traveller, also without sight, to go straight to the target, although presumably only so long as it is possible to keep to the original facing posture during locomotion.

Views from starting positions also involve the integration of visual and body-centred information. As noted earlier, the visual system works in conjunction with the vestibular system and proprioceptive inputs from head, neck and eye muscles that determine a person's stance relative to the goal. There is a close link between the eye and foot through muscular proprioceptive messages in the control of human posture, in which extra-ocular proprioceptive inputs seem to be dominant. Here again, the important factor seems to be the coordination of proprioceptive body-centred information with sight of the external scene prior to moving. Farrell and Thomson (1998) suggested that this coordination enables them to update their position later.

REFERENCE INFORMATION IN URBAN ENVIRONMENTS

Urban environments that prevent direct sight of the goal present problems also for the sighted. People cannot look around corners or through doors to find their way in an unfamiliar house. Walking through unfamiliar streets in an effort to find an underground station that someone told them was quite near would get fully sighted people there only by chance. They would be more likely to lose their way completely without more exact information. In principle, that is provided by descriptions of a new route, preferably with the number of side streets that need to be passed and with a landmark that indicates where to turn.

Route and landmark knowledge has been found easier for young sighted children than map information that provides an overview of the space to be traversed, and this has sometimes been suggested as a developmental sequence (Siegel & White, 1975). However, that is more a matter of familiarity and experience, and the complexity of the information, than a question of developmental stages. If the target is first presented in map form, the young can locate it later on a map quite successfully (Slator, 1982). Route information and landmark information also involve memory and knowledge. To be able to follow landmark and route instructions, the traveller, of whatever age, has to remember to apply the information first to his or her own position in space and to remember and follow the sequence of instructions. In principle, that is easier in graphic form, provided that the map reader is made aware of the meaning of the symbols and can align the map direction to his or her own position.

Vision provides potentially more useful reference information for landmarks than do sounds. Stationary objects in unfamiliar environments rarely emit meaningful "natural" sounds that can be substituted for visual reference and updating cues (Strelow & Brabyn, 1982). However, training in the use of low-frequency ambient sounds is likely to be particularly useful for the blind (Ashmead et al., 1998).

Some blind children discover spontaneously how to use echolocation. Hand clapping, stamping the ground audibly with the feet and, later, tapping the long cane produce sounds that blind children can use to detect their own distance from a wall and gaps in it. But they benefit from good mobility training and assisted learning by knowledgeable parents who realise the importance of providing all possible cues from external sources for reference and are able to link that information to the child's body-centred information and spontaneous strategies (Millar, 1994).

Complex spatial tasks that ask the pedestrian to reach a goal by an indirect path that has to be inferred from current information involve some form of mental work, whether by blind people or by sighted people, almost by definition. Hollins and Kelley (1988) showed that blind adults can update their position as well as the sighted. Nowadays, educated blind adults would also have sufficient knowledge about external layouts to be able to apply updating rules to target sounds when necessary. However, the fact that a given task produces roughly similar results does not necessarily mean that the covert mental work for the two tasks is identical.

It is often forgotten that problems which involve the reversal, rotation or updating of spatial locations do not necessarily require mental rotation by reference to an imagined external background. In principle, it is perfectly possible to solve posture reversal or rotation by using geometric rules to update a current body posture. Applying a left–right reversal rule, for instance, changes your body posture by 180 degrees from the initial facing position. Mobility training usually emphasises the importance of executing right-angled turns that lead to some intermediate haptic landmark, such as the edge of a path in walking across an otherwise open space. Young blind children also experience the difference in sound from a known source when they face it, and when they turn their back to it. But in environments that provide no other updating cues, it may be necessary to alert people, at least initially, to integrating the difference in sound with the required turns of the body.

Most congenitally totally blind children require some assistance or relevant feedback from experience initially in learning how to apply apparently simple geometric rules. It is necessary to distinguish such problems from spatial difficulties due to genetic dysfunction of relevant cortical (prefrontal-parietal) circuits. Children born with "Williams" syndrome typically perform poorly on visuo-spatial tasks. But these deficits do not correlate significantly with concomitant visual impairments (Atkinson, Anker, Braddick, Nokes, & Mason, 2001; Atkinson et al., 2003). Given the evidence reviewed in Chapter 2, one example suffices to show that the spatial difficulties of blind children are not necessarily due to a lack of spatial ability.

I used a tactile compass with braille letters indicating the geographical directions to test the application of a simple left–right reversal rule. An extremely intelligent young blind child learned that rule very easily indeed. In tests, she made no mistake in turning her body in the directions that were indicated to her by the letters on the compass. However, on leaving the testing room she turned left into the corridor, instead of turning right to go back to her classroom. In other words, she repeated the left turn she had made to enter the room, instead of reversing it and turning in the opposite direction when leaving. She thus relied on memory of her previous turning movement rather than on the recently learned rule (Millar, 1994). Such mistakes occurred much less when leaving a familiar classroom to turn into well-known corridors. Even when habitual movements need to be updated, familiar spaces afford known tactual landmarks for updating movements when necessary. With feedback from errors, and/or being made aware of the implications of a rule for orienting movements, blind children come to apply such rules successfully. The point is, rather, that coordinating locomotion with the use of updating rules is not necessarily automatic.

The mistake was not due to blindness. Occasionally turning in the wrong direction is hardly an unknown error by the sighted. Using movement direction or other body-centred cues is shown more by people with little or no visual experience (see Chapter 2). At the same time, individuals differ widely in the experiences they encounter, and in what their elders and peers expect them to attend to at different periods in time. That difference is generally recognised, but rarely applied. Data on errors in blindness are still treated too often as if they did – or, indeed, could – demonstrate "incapacity" rather than as pointers to questions about rehabilitation.

But the example also speaks to the more general question about the integration of different types of information in spatial tasks, which is considered next.

LOCOMOTION IN LARGE-SCALE SPACES WITHOUT STABLE REFERENCE CUES

Normally sighted people typically get lost in the desert, swimming under water or walking blind in unfamiliar spaces or in dense fog (Lund, 1930; Ross, 1974). What all such conditions have in common is that the environments lack easily identified or predictable stimuli that could potentially indicate a goal for the traveller or act as intermediate landmarks that could update movements as prescribed in written or graphic representations of the route.

Interestingly enough, people's trajectories through environments that lack distinctive features cannot be described in terms of chance performance, or by random movements that simply happen to occur by chance. Typically, people veer from the potentially straight-ahead path consistently in a particular direction. The deviation gets larger with distance, so much so that the trajectories give the impression that they will eventually end up at the starting point. That probably prompted the original explanation that animals, including man, have an innate, biologically useful spiralling mechanism. The idea is that veering by an animal that is lost functions to return it eventually to its original starting point (Guldberg, 1897; Schaeffer, 1928). However, although people who are lost in fog or mist often report that they walk in circles, there are no actual empirical records of complete circular trajectories for any animal, as far as I know, nor any concomitant evidence for a specific circling mechanism in the brain.

The tendency to veer consistently has been amply documented for blind as well as sighted people (Cratty, 1971; Guldberg, 1897; Guth & LaDuke, 1994, 1995; Lund, 1930; Ross, 1974; Rouse & Worchel, 1955; Schaefer, 1928). But findings and explanations differ. Veering predominantly to the right side is reported in some studies (Claparède, 1943; Rouse & Worchel, 1955), but by no means in all studies. The explanation that veering relates to cerebral dominance and handedness seems unlikely. Most people are right-handed, but at least one study found more veering to the left. Moreover, there is far more variability in the direction of veering by individuals than would be expected from the laterality hypothesis by the predominantly right-handed adults that normally participate in such tests (e.g. Guth & LaDuke, 1995).

Structural asymmetries of the body, such as differences in the length of the legs, or imbalance in kinaesthetic cues to body-posture (Lund, 1930; Ross, 1974; Rouse & Worchel, 1955) have also been used to explain veering. Lund (1930) tested over a hundred blindfolded participants on a longish (300-ft) route across a football field. Interestingly, unlike participants in a study mentioned earlier, they evidently veered from the straight path to the goal, despite having been shown the visual target prior to being blindfolded. It is possible that participants failed to orient their stance to the visual goal before being blindfolded. But that was not reported. The main result being quoted was that a high percentage of people whose left leg was longer than the right leg veered to the right when walking across the field, and they also did so when walking back to the starting point. However, the reported difference in leg length (5 mm) leaves room for doubt about the accuracy of measurements (Howard & Templeton, 1966). Indeed, other studies did not find significant correlations between structural asymmetries of the body and the direction of veering, although brief interpolated movements produce errors (e.g. Cratty, 1971; May & Klatzky, 2000). Movement effects may relate to the suggestion that veering could be due to an asymmetry in the vestibular system that would affect posture (Howard & Templeton, 1966).

Extensive studies by Cratty (1967, 1971) and Cratty and Williams (1966) showed that external cues from the floors of different test spaces affected the direction of veering. The external cues in this case were differences in the gradient and textures of the floors in the different spaces that participants could feel, or possibly hear from the sound of footfalls. The blind tended to use such cues more, as would be expected from their experience or mobility training. However, given the importance of sounds for guiding blind locomotion, the finding that external noise had no effect on veering (Rouse & Worchel, 1955) is surprising. There could be more than one reason why the presence or absence of external noise produced no difference in veering in that study. Measuring movement biases in large-scale space precisely presents a significant problem for studies of veering.

Video-recording participants' walk across a field, by following behind them with the camera (Guth & LaDuke, 1995), provided greater accuracy than earlier methods. The study was designed to test individual differences with four blind female pedestrians. The ground was marked with a visible vertical (north–south) middle line, crossed at 5-m intervals by horizontal lines.

Video replay of the tapes made it possible for observers to judge the distance of a participant's body midline from the field midline as she crossed a horizontal line. The participants wore earphones linked to a special transmitter system that was located at the (north) starting point of the walk. They could hear environmental noises and instructions, but could not localise them. That may have applied less to noises that came from the cameraman following from behind, especially in the early part of the route.

The interest was in individual differences in veering by blind pedestrians. The findings, calculated in terms of algebraic errors (signed positive for right, negative for left veers), showed differences not only between the four participants, but considerable variability also for each individual across three testing sessions. Measured from the last line crossed, one woman veered left in all three sessions, but her deviations increased over the sessions. One woman veered to the right in all three sessions, with a slight decrease in deviations over the sessions. The other two women veered to the right in the first session, and to the left in the last two sessions. The deviations by one of them were relatively small. The other deviated most in the middle session. However, despite the variability both between individuals and over sessions, each path as depicted seemed to continue in roughly the same general direction.

Previous studies thus established that blind and sighted people veer consistently from the direct path. The direct path is the shortest route from the starting location to the opposite one facing it, when walking through spaces that afford no cues for target locations and no recognisable landmarks that could update locomotion. The direction of veering to the right or left seems to vary. However, once deviations from a path have started, they are not random, but seem to continue more or less in the same direction.

It was much less clear what actually produces veering in the first place. Previous studies left an uncertainty about the effect of external cues on the direction of veering. The finding that cues from floor textures, and possibly from the sound of footfalls, related to directions of walking blind or blindfolded was positive. But it seemed to be contradicted by the finding that external noise was unrelated to the direction of veering. It is not obvious why. The question of whether or not unexpected extraneous sounds have effects on veering is important for practical, as well as theoretical, reasons. But it needed testing by conditions that varied the location of an unexpected sound systematically relative to the movement path.

If anything, the previous evidence on effects of asymmetric body structures seemed even more dubious. The findings on movement effects suggested a possible role of movements in altering posture cues. Indeed, effects of biomechanical posture cues on locomotion would be expected from evidence in normal environments. But previous findings had been based on cumbersome and somewhat imprecise measures of asymmetries in body structures, such as leg length. These are bound to vary widely between people of different height and weight. The evidence was thus not decisive. It needed to be checked out in conditions that varied posture cues experimentally.

A METHOD AND RESULTS ON VEERING BY BLIND CHILDREN

Spaces that lack stable cues that can be integrated for reference provide a rather powerful test of the reference hypothesis. The hypothesis – that spatial processing depends on integrating inputs from the available sources as reference cues for the task to be performed – implies that accurate reference organisation depends on the convergence and partial overlap of external cues, such as sights or sounds, with body-centred proprioceptive and posture cues. Moreover, it assumes that conditions in which cues from different sources do not converge or overlap would fragment reference organisations and bias locomotion accordingly.

The hypothesis predicts that the very cues which normally combine to produce accurate performance in large-scale space bias locomotion systematically when they occur in isolation in spaces that lack reference cues that are potentially convergent. Previous findings suggest that accurate blind performance in large-scale space depends on the coordination of inputs from sounds with inputs from body-centred posture cues. The relevant inputs to be tested in isolation, therefore, were sounds and conditions that change body-centred posture cues.

The hypothesis was therefore that unexpected, irrelevant, isolated sounds would be more likely to elicit orienting reactions towards the source of the sound. It would be difficult to coordinate an irrelevant sound that occurs after locomotion has started with posture cues adopted earlier, so as to provide reference information that would maintain the direction of locomotion.

The hypothesis also implies that biomechanical body-posture cues bias the direction of locomotion when there are no external cues that could be integrated with inputs from the body. Previous findings on the effects of asymmetric biomechanical or posture cues had been based on very imprecise measures. These were ruled out as impracticable as well as necessarily inexact.

Experimental manipulation of postures, rather than imprecise measures, was needed. Postures can be altered. When people carry a weight with one hand, they lean to the opposite side for balance. Posture cues were therefore manipulated experimentally by getting participants to carry a weight with either the right hand or the left hand. The prediction was that carrying the weight in the right hand would produce veering to the left, and carrying the weight in the left hand would produce veering to the right.

The method and findings of two studies that tested the predictions (Millar, 1999b) are described here in some detail because the findings were based on a very precise measuring and recording system. More to the point, the measuring system also produced some unpredicted but potentially important effects.

To test the predictions from the reference hypothesis, a sensitive automatic recording and calculating system was needed that would measure any deviation from the straight-ahead continuously and precisely from the start. Precise, continuous data on locomotion from the start were needed to test how a subsequent irrelevant external sound affected the direction of veering. A system that would measure any deviation from the straight-ahead continuously and precisely was needed also to enable effects of experimentally changed postures to be recorded and calculated automatically for early as well as later stages of locomotion.

The design of the apparatus used in the studies was adapted from the monitoring system that Brabyn (Brabyn, 1978; Brabyn & Strelow, 1977; Strelow, Brabyn, & Clark, 1976) devised. Our system had to be fully transportable. It needed a relatively small experimental space that could be set up in any largish, quiet, unfamiliar, carpeted room in a house set somewhat apart in a building complex that was convenient for testing blind children.

An experimental (2 × 2 m) space was constructed with four corner posts (A, B, C, D, Figure 3.1). They were anchored in heavy metal bases so that they could not be displaced easily by accidental contact. The posts carried four transverse wooden safety rods for the sides of the square at a height of 60 cm. The point of testing in a carpeted room was that it prevented noise from the floor and echo from the walls of the room during locomotion.

The device was designed to record locomotion continuously (every 70 ms) from the starting point. Three heavy metal stands, each carrying a potentiometer, were placed in an equilateral triangle at the midpoint of one (A–B) side and at the opposite (C, D) corners of the experimental space. A weighted pulley (take-up spool) at the top of each stand extended or contracted a light line that was fastened to the end-tip of a lightweight metal rod on a backpack worn by the participant. The backpack consisted of a comfortably padded, light but stable, rigid frame. The metal rod extended upwards from the backpack to above the participant's head. The top of the rod was bent at right angles, so that the end-tip with the three lines was located directly over the mid-axis (Z-axis) of the participant's head and body (Figure 3.1). The stands were constructed to extend in height (from 3.1 ft to 6 ft). They were adjusted to the height of the end-tip of the rod above a participant's head.

The system recorded and output data automatically. It included a power supply, modulator, demodulator and computer interface. The voltages across potentiometers were sampled at 70-ms intervals. They were transformed to FM audio-frequency signals. The outputs were to a tape recorder. The tape recordings were decoded in real-time by hardware electronics. Each sample of a participant's position was output as an event, with the structure shown in

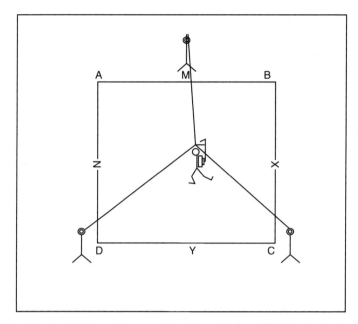

Figure 3.1 Sketch of the layout (not to scale) of the test space A-B-C-D, showing the position of the three stands carrying the potentiometers and pulleys for the uptake of the lines attached to the tip of the rod on the harness, which extends to the midline (z-axis) above the participant's head and body. The 8 routes were as follows: A–C, C–A, B–D, D–B, M–Y, Y–M, X–Z, Z–X. [From S. Millar, "Veering re-visited: Noise and posture cues in walking without sight", *Perception*, 1999, *28*, 765–780. Reproduced with permission from the publisher, Pion Ltd., London.]

Figure 3.2. Each potentiometer signal had a definite time relationship with the reference signal at a fixed frequency and sequence.

The signals were scanned repetitively. The composite signal was fed into the demodulators and reconverted the frequencies into voltages that were identical and in the same sequences as the potentiometer voltages, and gated appropriately. The computer interface produced computer-compatible outputs from the demodulator outputs and converted analogue to digital outputs from the decoder. A software programme converted all outputs into X–Y coordinate sequential positions.

The computer output data were digital, in terms of coordinates of the successive sampling of all locations in each trial. The output thus gave the sequential positions occupied during locomotion from the start to stop, in Euclidean (horizontal–vertical) coordinates in either digital or graphic form. The main data for statistical analyses were (algebraically summed) deviations to the right (signed as positive) and to the left (signed as negative) from the straight-ahead path from the origin to the participant's stopping point. The stopping point was determined by the experimenter's saying "stop" (at a point

3. REFERENCE CUES IN LARGE-SCALE SPACE   57

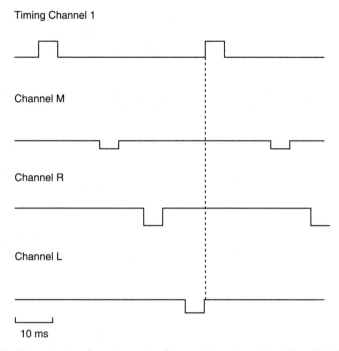

Figure 3.2 Event structure for each sample of the participant's position. [From S. Millar "Veering re-visited: Noise and posture cues in walking without sight", *Perception*, 1999, *28*, 765–780. Reproduced with permission from the publisher, Pion Ltd., London.]

approximately 10 cm) before a participant's body would otherwise have touched the barrier around the experimental space. Participants did not encounter the barrier. They received no feedback on any trial.

Nine blind children took part in the study on sound effects. They ranged in age from 8 to 10½ years, and from low average to very superior intelligence on test. Medical records showed that eight children had been totally blind from birth. One 8 year old who was totally blinded at the age of 3½ was included. As noted earlier, none of the children deviated from the straight-ahead in walking (2 m) to the source of the sound of a well-known voice, either immediately or after a short delay in the baseline control condition.

The children were all keen to play "being spacemen" and to comply with the injunction to try to walk in as straight a line as possible on the "space-walks". The experimental space had been set up in a room that was unfamiliar to the children, in a quiet house set apart from other buildings. None of the children deviated from the straight-ahead in walking 2 m to the source of the sound of a well-known voice, either immediately or after a short delay. A thick carpet prevented echo or noise from the floor and walls of the room during locomotion. There was no noise from outside the window. In addition the test routes in the experimental space were started from different, counterbalanced locations. The order of test conditions and routes was randomised across participants, with the proviso that experimental and control conditions should be tested roughly equally for every route.

The point was to simulate a space that did not provide external frame information. The children were, therefore, not led around the square space at any time. Care was taken also that they did not touch the surrounding barrier from inside the space at any time. The children were tested singly. The experimenter guided the child to the starting point of each test route, helped the child to assume a straight posture, and ensured that the head, body and feet postures were aligned to face straight-ahead at the start of the walk.

The walk began at the signal "Go", given directly behind the head of the child. The experimenter moved noiselessly to a random location in the "no-noise" (N) control condition; to the barrier on the right of the participant for the "noise to the right" (R) condition; and to the left in the "noise to the left" (L) condition. The sound in the noise condition consisted of two short bursts from a muffled bell, located outside the experimental space, parallel to the route, at a distance of 0.5 m from the start.

The noise was designed to be completely irrelevant to the game and was not expected by the participants. The results showed that the children veered to the right when the sound was on the right and veered to the left when the sound was on the left, as predicted by the reference hypothesis (Figure 3.3). Algebraic errors of degrees of deviation from the straight-ahead were calculated from the start of the route to the stopping point. Errors were signed positive for deviations to the right (R) and negative for deviations to the left (L). The variation between individuals was high, as found in most previous studies. But the difference between right and left deviations was significant. The graph (Figure 3.3) shows that, on average, there was also a (positive) deviation to the right in the control (N) condition. However, the degree of veering to the right was significantly larger in the R than in the N condition. Veering to the left when the noise was on the left (L condition) also differed significantly from the control (N) condition. The mean error in the L condition was based on eight rather than nine participants, because one child veered so far to the left in the L condition that she had to be stopped. The incomplete data had to be omitted from the statistical analysis. But they actually showed the largest deviation to the left in the "noise-left" (L) condition (Figure 3.3).

The findings showed for the first time that locomotion veers significantly in the direction from which an irrelevant sound emanates. They were consistent with the prediction that an unexpected, isolated irrelevant external sound affects locomotion significantly. Although the children were blind, some of the young participants were extremely bright. In any case, sighted adults and children also veer from the straight-ahead in spaces that afford no reference cues. Knowing geometry or being able to mentally represent spatial layouts

3. REFERENCE CUES IN LARGE-SCALE SPACE   59

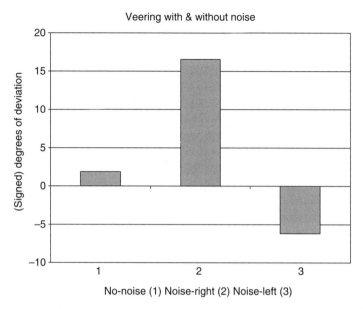

Figure 3.3 Mean degrees of deviations to the right (signed +) and to the left (signed −) in conditions N (no-noise control), R (noise on the right) and L (noise on the left).

does not help in such conditions. Consistent with previous findings, once veering has started, the trajectories continued more or less along the deviated paths.

The examples in Figure 3.4a–d of trajectories by congenitally totally blind children are reduced copies of actual graphs from digital outputs, superimposed here to compare two conditions that are of special interest. Figure 3.4a is an example of a trajectory by a boy in the N (no-noise control) condition, showing minimal veering to the right and left, but still keeping fairly closely to the straight-ahead, compared to his trajectory in the L (noise-left) condition, which starts to diverge more to the left after the noise (at 0.5 m noise) and ends by veering more than a metre to the left in the L condition. Figure 3.4b is the example mentioned above of two trajectories by the same child in two N (no-noise control) conditions, veering to the right with shallow steps on one (C–A) route, and much further to the right with wide sidesteps on another (Z–X) route. Figure 3.4c compares the N (C–A, no-noise) trajectory by that child with her trajectory in the L (noise-left) in condition on the X–Z route, where she veered so far left that she had to be stopped before reaching the barrier. Figure 3.4d shows the trajectories of a rather bright little girl in the N (no-noise control) condition, keeping mainly to the straight-ahead though veering marginally right from a marginal left start on an A–C route, but veering significantly further to the right in the R (noise-right) condition.

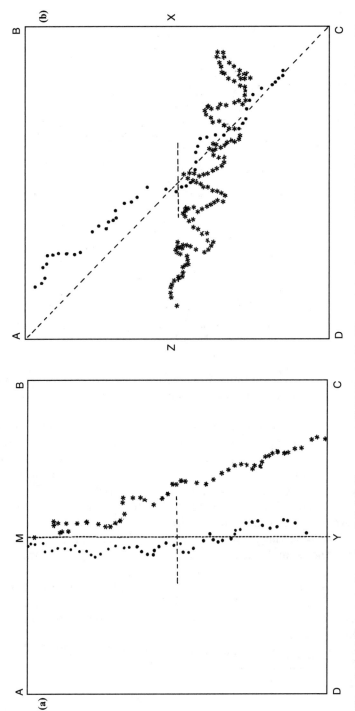

Figure 3.4 (a): Trajectories by a congenitally blind boy on route **Y–M** (**Y** origin) in **N** (no-noise) and on route **M–Y** (**M** origin) in the **L** (noise-left) conditions. (b): Trajectories by a congenitally blind girl in two **N** (no-noise control) conditions on routes **C–A** (**C** origin) and **Z–X** (**Z** origin), respectively. Closely packed, deep side-steps suggest slow progress. Copy of graphs of digital output (10 cm = 1 m).

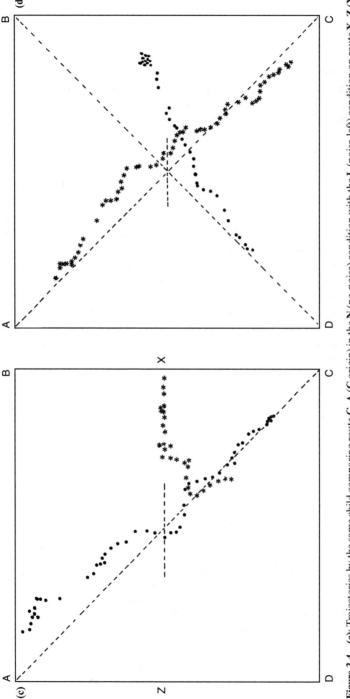

Figure 3.4 **(c)**: Trajectories by the same child comparing route **C–A** (**C** origin) in the **N** (no-noise) condition with the **L** (noise-left) condition on route **X–Z** (**X** origin) veering left, terminated before the barrier. **(d)**: Trajectories by another congenitally blind girl comparing the **N** (no-noise) condition on route **A–C** (**A** origin) with the **R** (noise-right) condition on route **D–B** (**D** origin), showing a right veer. [From S. Millar, "Veering re-visited: Noise and posture cues in walking without sight", *Perception*, 1999, *28*, 765–780. Reproduced with permission from the publisher, Pion Ltd., London.]

New, rather surprising evidence on detailed body movements was revealed by the frequency (70 ms) with which the system sampled sequential positions in trajectories. Underlying the main direction of the trajectories are small concave and convex body movements that oscillate from side to side as the participant steps forward alternately with the right and left foot. The difference in depth of the alternate concave and convex movements in two control conditions by the same child (Figure 3.4b) was noteworthy. The data points are much more closely clustered for the wide sidesteps in the Z–X control condition than for the shallow sidesteps in the C–A control condition. It suggested that the width of sideways steps varied with the speed of walking, wide steps being slower. This point is discussed together with findings in the second study.

The second study tested posture cues instead of noise cues. It was carried out three months later with fourteen (7 female, 7 male) blind children. Eight had taken part in the previous study. Of the six new participants with the same mean age and the same slightly above average intelligence, five were totally blind from birth, although one had a pupil reaction to light shone directly into the eye. The sixth child had been blinded totally at the age of 2 years but wanted to be included. All knew the experimenter well. The apparatus, experimental space, test routes and all procedures were precisely the same as in the previous study. Each child was tested singly in three experimental conditions. Conditions and routes were randomised over participants as before.

Asymmetric postures resulted from getting the child to carry a (2 lb) bag by the handle. Carrying a "load" in the right hand (R condition) made the child lean to the left for balance. Carrying the load in the left hand (L condition) made the child lean to the right. The no-load (N) condition was the control. In calculating constant errors, veering to the right was signed positive, and veering to the left was signed negative. Mean degrees of deviation from the straight-ahead were calculated from the midpoint of the straight-ahead to the stop, as well as from the origin of the route. The point was to assess if the average angle of veering (for this larger sample) increased or decreased further along the walk.

The results are graphed in Figure 3.5. When participants carried the load in the right hand (R), they leaned left for balance. They also veered to the left. Similarly, when participants carried the load in their left hand (L), they leaned to the right for balance and veered to the right from the straight path. Variability between and within individuals was high, as in all previous studies on veering (e.g. Guth & LaDuke, 1995). But the difference between veering left in R conditions and veering right in L conditions was significant, as calculated both from the starting point and from the midpoint of the walk. The mean deviations from the straight-ahead were larger from the midpoint (Figure 3.5), suggesting that the trajectories tended to deviate more and/or were less variable from that point on.

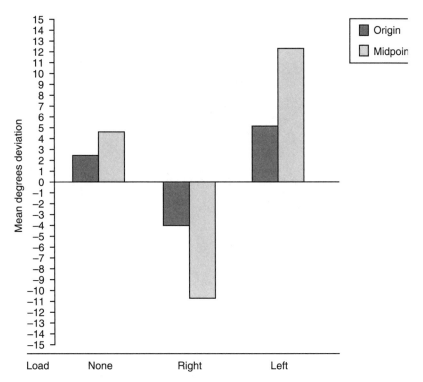

Figure 3.5 Mean degrees of deviation to the left or right in N (no-load), R (load on the right) and L (load on the left) conditions, calculated from the start (origin) and from the midpoint of the walk to the endpoint. [From S. Millar "Veering re-visited: Noise and posture cues in walking without sight", *Perception*, 1999, *28*, 765–780. Reproduced with permission from the publisher, Pion Ltd., London.]

The findings showed that experimentally induced asymmetries in body postures do produce significant differences in the direction of veering. They were consistent with the assumption that the very cues that normally contribute to accurate locomotion produce biases in conditions that fail to provide converging external cues that could be coordinated for reference or updating. Biomechanical cues from asymmetries in posture cues must, consequently, be considered a biasing factor in the direction of veering.

The computer-generated graphs from digital outputs for sequential positions in the trajectories again showed the alternating sidesteps that underlie progress in a given direction. These graphs also suggested that relatively deep convex and concave step-wise curves were associated with denser sampling of data for slow walks, and that shallow step-wise curves were associated with relatively sparser data sampling for faster walks. Moreover, the wider (deeper) curves underlying slower walks also seemed to produce more veering in experimental conditions.

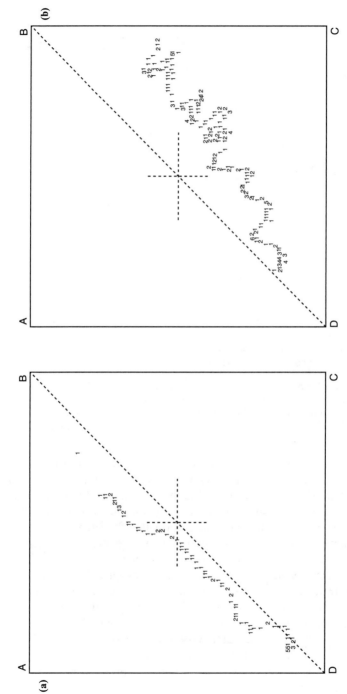

Figure 3.6 Examples of trajectories showing **(a)**: shallow sidesteps and spaced samples on route **D–B** (**D** origin) veering left (load-right); **(b)**: deep sidesteps and bunched samples on route **D–B** (**D** origin) veering right (load-left).

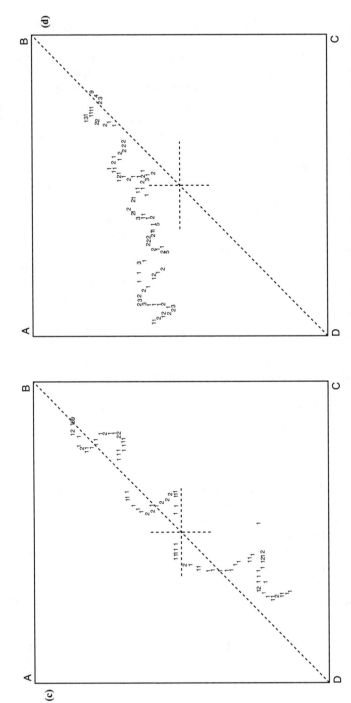

Figure 3.6 (**c**): deep sidesteps on route **B**–**D** (**B** origin) veering left (load-right); (**d**): deep sidesteps on route **B**–**D** (**B** origin) veering right (load-left). (From S. Millar "Veering re-visited: Noise and posture cues in walking without sight", *Perception*, 1999, *28*, 765–780. Reproduced with permission from the publisher, Pion Ltd., London.]

Four examples of actual trajectories by congenitally totally blind children are shown in Figure 3.6a–d. The digits on the graphs show the number of valid samples at each position. The first (Figure 3.6a) is an example of veering to the left with relatively shallow steps on a D–B route by a participant when carrying the bag with the right hand, and a significant but relatively small angle of veering. Example 3.6b is a right-veering trajectory for route D–B with an L load, showing closely bunched position samples for every alternate sidestep, and ending far to the right before being stopped as getting close to the barrier. Example 3.6c shows a left veer with the bag being carried by the right hand on the B–D trajectory with deep underlying sidesteps, with bunched data (one far to the left of the path). Example 3.6d is a B–D trajectory veering to the right with a load on the left, showing 1-m veering to the right, with closely bunched data for the sidestep curves. Some positions attracted more data points than was possible to accommodate in the small copies. They are therefore shown by digits that indicate additional numbers of data points.

The new data on the alternating side-to-side steps that underlie forward locomotion in a particular direction suggest that these movements indicate additional factors in veering that have to be taken into account. Two possibilities are now considered.

Effects of laterality seemed highly unlikely as a major factor. As noted earlier, veering to the right and left varies far more, not only between individuals, but also by the same people, for that to be the case. Moreover, the same variability was shown by even the most right-sided children in a pilot study. The pre-test study had been conducted with ten blindfolded sighted 8-year-olds in order to check out the system, test the apparatus, the harness and the test routes and find a suitable "load" that would produce asymmetric postures. Six of the children were right-sided on four subsequent laterality tests (right eye for looking through a small hole, right hand for drawing and writing, right foot for kicking). Three of these children veered right, one veered left and the other two veered in both directions in two control conditions. Most of the blind children were also right-handed. Individual participants in the two studies discussed here showed a similar variability in veering left or right in neutral conditions, as reported by previous studies.

The fact that the average deviation in control conditions was positive (to the right) needs to be considered briefly, although it could not explain the direction of veering in the relevant experimental conditions. But the possibility of some additional effect from some "natural" biomechanical imbalance cannot be ruled out. A tendency to veer to the right might reduce veering in the direction, or in the opposite direction, of experimentally produced "incidental" sound and body-centred cues. But if so, it evidently did not override the significant effects of veering to the left with experimentally manipulated cues. Forms of structural imbalance would vary too widely between individuals to be established easily in the first place and would also increase variability between participants. In any case, asymmetries in individual body-structures could, at most, enhance or diminish effects of other isolated or discrepant cues. They could not explain why the same individual can veer in different directions on different walks, which almost all studies report and was also found here in neutral conditions, in some cases for the same walk, by the same participant.

The new finding, that alternating convex and concave curves underlie all forward trajectories, is important. Differences in depth or shallowness of the step-wise curves underlying different trajectories could not be attributed to an individual's characteristic locomotion. The same participant produced a pattern of deep-bunched sidesteps in one control condition and more spaced shallow sidesteps in another control condition (Figure 3.4c). The difference in the depth of step-wise curves also occurred for same route in different experimental conditions (Figures 3.6a–d). However, deep sidestep curves were associated with increased numbers of data points or clusters at sequential positions at the (70-ms) sampling rate, implying relatively slow walking. Shallow side-steps underlying a trajectory seemed to be associated with sparser data points at sequential positions, sampled at the same (70-ms) rate, implying faster walking.

The data are relevant to the widely held belief that fast walking reduces the degree of veering, whereas slow walking increases veering (Welsh & Blash, 1980). Cratty and Williams (1966) found a significant negative correlation between veering and the speed of walking for 100 m by blind people, although the correlation was no longer significant when measured at 200 m from the starting point. When instructed to walk faster or more slowly than their normal walking speed for a 14 m distance, a majority of blindfolded sighted participants veered less when walking faster (Klatzky, 1999; Klatzky et al., 1990). The finding that highly anxious blind pedestrians walk more slowly (Harris, 1967) is of practical importance. But the many reasons why people sometimes walk more slowly than usual do not, as such, explain why walking speed should influence deviations from the straight-ahead.

The finding on deep and narrow steps from side to side that underlie forward trajectories are of interest in that regard. The point is that the width and length of sidesteps affect body postures. As noted earlier, to walk straight-ahead in a space that affords no reference or updating cues, it is essential to maintain the facing posture during locomotion that was adopted at the start. Deep convex and concave curves for sidesteps with small forward moves would make it more difficult to maintain or return to the initial facing posture than long, narrow sidesteps. The frequent sampling rate of the recording system produced more data points for the sequential positions for smaller forward moves with deep sidesteps than for the longer forward moves with shallow sidesteps. Locomotion is thus slower in trajectories with underlying deep sidesteps than with underlying shallow sidesteps. A slow wide step to the side would produce a greater deviation from the initial facing posture and also longer-lasting proprioceptive cues that could interfere with the proprioceptive and movement cues needed for righting it and resuming the initial facing posture.

The association between walking speed and the extent of veering from the straight-ahead thus makes good sense when considered in terms of the compatibility of proprioceptive and movement cues involved in body-postures and in side-to-side and forward movements. It also explains why slow walking by anxious blind people produces more veering. Fast walking is hazardous if you are afraid or uncertain for any reason. Slow, wide sidesteps prevent stumbling and violent impacts from bumping into obstacles. People are not usually aware of veering, much less that slow walking may increase it. Initial deviations may occur with the first sidestep that alters the facing position in control conditions without other biasing cues. It is possible that for some individuals – though clearly not for all – the "dominant" leg influences the initial direction of veering. If so, it would explain why small, variable veers to the right rather than to the left are sometimes found for control conditions. That is speculative. Individual differences can only be checked out individually, which may be needed for practical purposes. The main point is that the unexpected findings provide new information on the lateral movements that underlie forward trajectories.

The new findings support the assumption that same cues that produce accurate locomotion when they converge and overlap explain veering when they occur in isolation. An irrelevant, isolated external sound to the left or right changed the direction of locomotion. Carrying a load on one side produced deviations to the opposite side. Slow, deep sidesteps affect facing postures more and for longer than long, narrow steps if they are isolated from correcting reference cues that would restore initial facing postures. The factors that produce the reference organisation that makes locomotion accurate bias locomotion when they cannot be integrated.

SPATIAL KNOWLEDGE, GEOMETRIC INFERENCE AND PERCEPTUAL CUES

This chapter started with the question of what the traveller contributes to solving the different tasks that large-scale environments pose. The findings described earlier suggest that solutions depend largely on the perceptual information that different spaces afford the traveller to reach his or her goal. In some respects, there seems to be an inverse relation between a paucity of current inputs and the knowledge and inferential procedures that the traveller needs to deploy. However, that is a partial factor at best. Not even a complete set of innate geometric principles would help the pedestrian in the total absence of perceptual cues. The point is not trivial.

The majority of blind children veered in the direction of the sound, and at least one turned directly towards it. Not one used the sound as a reference or guide for steering a straighter path. Nevertheless, the same congenitally totally blind children were able to walk the same distance straight-ahead without deviating on pre-test when they started with an established facing posture to a previously heard meaningful sound.

The fact that blind and sighted adults veer in much the same way in spaces that lack convergent reference cues suggests further that the potentially better geometry, more extensive knowledge of external spaces or ability to mentally represent spaces by adults does not prevent veering.

The notion that perceptual inputs can be relegated to some inferior theoretical position relative to geometric knowledge or to mental representations of spaces is thus not tenable. The reverse idea – that covert problem solving or spatial representation is unnecessary, at least in more complex spatial tasks – is equally untenable.

A study which showed that blind and blindfolded sighted adults were able to reach targets by indirect as well as direct paths (Loomis, Lippa, Klatzky, & Golledge, 2002) provides supporting evidence for both effects, although that was not its primary aim. Targets were specified verbally, such as "at 2 o'clock, at 16 feet" or by a sound. Verbal descriptions proved as good as sounds for getting to the target by an indirect path after having to turn from the initial path. But the interest here is in the experimental details, because indirect paths, like rotation tasks, involve covert planning and/or mental reorganisation. All participants had good travel skills. They were given some useful prior information and some practice in recognising sounds to the left or right. Even the mention of clock locations would be relevant to the task of imagining objects at target locations from different positions. Prior to direct and indirect walks, participants faced straight-ahead to a midfield sound initially. Facing postures are clearly disrupted by having to turn in a different direction, especially after the offset of target information. But it seems reasonable to suppose that the initial coordination of posture with target cues, combined with inferences from the other information, was a factor in enabling participants to construe target locations from indirect locations reasonably well. Perceptual information, memory and contextual knowledge complement each other, even if to a different extent in different tasks. They are not alternatives.

The evidence reviewed here supports theories, including the reference hypothesis, that stress the importance of interactions between inputs from different sources for reference information in spatial tasks (Gibson, 1966, 1979; Howard & Templeton, 1966; Paillard, 1991). A more recent study (Gugerty & Brooks, 2001) also suggests that integrating "heading" information – in this case, from vision – with egocentric reference cues is important for accurate judgements of cardinal directions.

The findings also support the further implication that a single, unfamiliar, irrelevant sound or body-centred input is more likely to bias locomotion than to serve as an adequate updating or reference cue. At least two converging inputs from external and body-centred sources evidently need to be integrated for the integrating activity that produces relatively accurate reference and updating cues in spatial tasks.

CHAPTER FOUR

Hand movements and spatial reference in shapes and small-scale space

Traditionally, movements have been studied mainly in terms of their output functions and their control by inputs from vision and other sensory sources in acting on the environment, and also in terms of feedback for corrective action and of feed-forward information in planning actions before they are executed. These systems are relevant to actions in small-scale space as well as in large-scale space.

However, the focus in this chapter is on finger and hand movements as important sources of input information, together with touch, in small-scale tabletop space. The question of how finger and hand movements relate to spatial coding is considered first with respect to the tiny raised-dot patterns that constitute braille characters, and then with regard to larger raised-line spatial layouts.

How spatial coding by touch and movement takes place with these materials is of considerable interest, not least because both braille and raised-line and embossed spatial layouts are the two most important sources of linguistic and spatial information in blind conditions.

Findings on braille are considered first. They exemplify how apparently trivial differences in the physical composition and size of haptic materials influence finger and hand movements and, in turn, the perceptual features that the combination of movements and touch pick up with experience of the materials and experience of task requirements. Findings showing that – and why – single braille patterns are less easily coded as "global" shapes in touch are reviewed first.

Lateral scanning and spatial coding of tactile patterns are considered in the next section. Findings which suggest that lateral "dynamic" shear patterns, rather than the pick-up of global letter-shapes, characterises fast text reading are examined. The evidence suggests that the differentiation of scanning movements for different types of reading task depends on experience and proficiency and permits the pick-up of spatial as well as verbal information alternately by both the left and right hands.

The relation of the two hands to the contralateral cerebral hemispheres in spatial, movement and language processes is considered next. It is argued that not only verbal factors, but also the movement information that is involved make it impossible to infer spatial coding directly from a left-hand advantage, as has often been assumed. An experiment that was designed to test haptic spatial coding by the right and left hands in relation to added reference information is described in some detail.

The new evidence shows that providing explicit relevant reference cues makes it possible to infer spatial coding, regardless of hand advantages. Better performance with either hand could also depend on other factors that are lateralised differently. The method provides a consistent behavioural diagnostic tool of spatial coding that can be used independently of hand effects.

FINGER MOVEMENTS AND CODING BRAILLE CHARACTERS AS SHAPES

Katz (1925) was evidently right to suggest that global form is not the easiest basis for haptic perception of unfamiliar shapes, at least initially. His findings are consistent with more recent empirical evidence. There are a number of reasons why that should be so.

In some respects, perception by touch is a misnomer. The majority of tactile shape and spatial tasks in tabletop space require active exploring movements. Haptic perception thus involves a combination of inputs from touch and movement that seem much less tightly organised, or vary more with task conditions, than multisensory inputs to the visual system. Arm, hand and finger movements vary in type and sweep with practice as well as with task demands. Detecting the features of a new stimulus object involves sequential movements, depending on the size of the stimulus object and on the consequent concomitant or alternate deployment of fingers and of the two hands needed to explore them. The advantage is that touch and exploring movements convey more information than touch alone.

Types of exploratory movements matter (Davidson, 1972; Davidson, Appelle, & Haber, 1992). I remember my surprise as a young researcher to hear a competent braille teacher ask a child to stop constantly rubbing over a braille letter. He was to keep his finger still, otherwise he would not learn to recognise the shape of the letter. The injunction was initially puzzling. Prolonged touch, especially when pressing hard, as young beginners tend to do, results in loss of sensation, much as prolonged fixation does in vision. Moving the finger restores feeling. Active touch, which involves movement, tends to be more efficient than passive touch (e.g. Gibson, 1962; Heller, 1983, 1984a, 1984b). But the teacher had a point. The constant, indiscriminate rubbing over a single letter by beginners, familiar to braille teachers, brings no advantage.

Unsystematic scanning is not confined to touch. It has long been known that young sighted children scan unsystematically when asked to compare objects that differ only by a small visual feature (Vurpillot, 1976). There is a difference, nevertheless. In vision, information is taken in when the stimulus is being fixated rather than during saccadic eye movements. In haptic perception, by contrast, the main intake of information occurs during scanning (Bürklen, 1932; Davidson, Wiles-Kettleman, & Haber, 1980; Foulke, 1982). However, scanning needs to be systematic and adapted to the task, and that raises questions that need to be considered.

The teacher's advice actually implied that braille characters are, or should be, recognised most easily as global shapes. The assumption that braille characters, like print letters in vision, are best recognised holistically or as global outline shapes (Bürklen, 1932; Nolan & Kederis, 1969) is still made by many researchers, and by many braille teachers, even though the current emphasis has now turned to the recoding of letters. Fast recoding into phonological form is important (Baddeley, 1990). It is, indeed, even more important for short-term recall in touch (Millar, 1975b) than it is for visual materials in verbal tasks. But the assumption that recoding is necessarily based on outline shapes derives from what is supposed about the detection of visual print letters. This is understandable. Visually, braille characters do indeed look like quite simple shapes (Figure 4.1).

The fact that the physical composition of braille characters is not only quite different from print letters, and that this affects the features that are most easily picked up by relevant finger movements, is rarely taken into account, although it is quite well known that the characters all derive from one small (6.2-mm) matrix of six (2 × 3) raised dots. The presence or absence of any dot signifies a different character. Print letters are easy to recognise and to discriminate because they consist of different combinations of multiple features. Each print letter presents a different combination of straight vertical/horizontal and oblique lines and convex or concave curves or semicircles. The recognition rule is the same in touch and in vision. Print letters that differ least in features (b & d; p & q) are the most easily confused. Braille characters conspicuously lack the redundancy of different features that makes for the easy, fast identification of most print letters in vision. The shape of braille letters can, of course, be recognised with practice. But sole

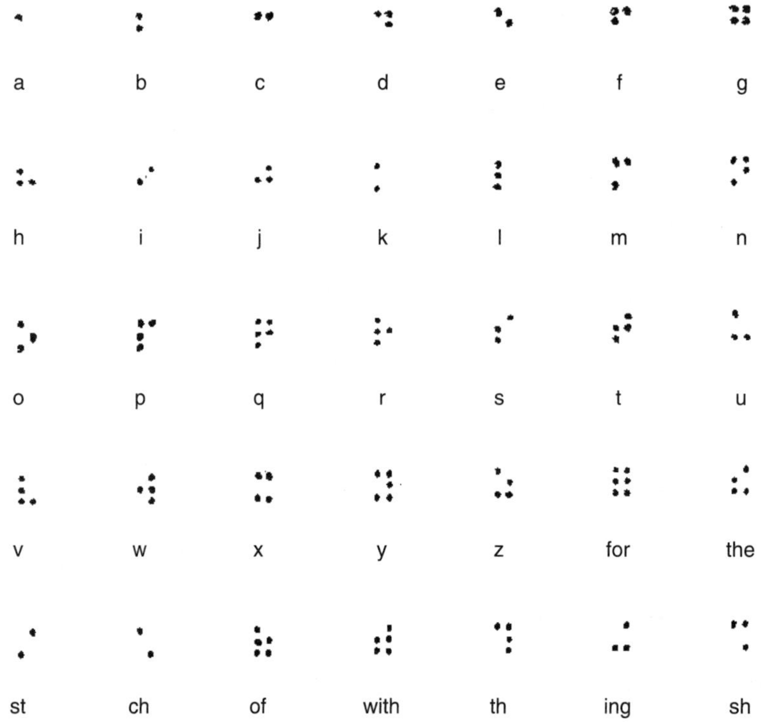

Figure 4.1 The braille alphabet and examples of some single-character contractions. [From S. Millar (1997), *Reading by Touch*. Published by Routledge (Taylor & Francis), London. Reproduced with permission.]

reliance on recognising braille characters as shapes depends on circular or up-and-down finger movements that are associated with slow letter-by-letter reading (see later).

Difficulties in learning braille are usually attributed to the small size of the characters. Tactile acuity matters. However, the fingertip is more sensitive than almost any other part of the skin except the lips (Mountcastle, 2005). Acuity, as measured by the distance at which two raised dots can be distinguished, is usually sufficient for two raised dots that are at a distance of 2 mm from each other, as they are in braille characters.

Long-term experience with braille produces specific advantages for braille-like dot patterns. Braille readers outperformed age-matched sighted controls in discriminating braille-like dot patterns, but not in discriminating gratings that differed in ridge width or in orientation of gratings (Grant, Thiagarajah, & Sathian, 2000), suggesting that the advantage is task-specific. Proficient braillists retain their ability well into old age (Stevens, Foulke, & Patterson, 1996).

The small size of the characters does make it impossible to code the location of the dots in a pattern spatially by reference to body-centred or body midline cues. But placing the forefinger gently, and not for too long, on a character makes it possible to use the top and sides of the finger as a literal spatial frame to which the location of dots in the character can be related. That strategy is sometimes preferred, and it is even advocated by former print readers, But if it is continued, it can lead to slow letter-by-letter reading. The main difficulty is that the system totally lacks the multidimensional redundancy that makes print letters easy to identify.

It was worth asking why Louis Braille, who was blind himself, disliked the raised print letters that he had to learn to recognise as a blind child and, instead, invented the non-redundant script based solely on the tiny six-dot braille cell. Braille's choice of a small six-dot matrix from which all characters derive, starting with one dot for "a", two dots for "b", and so on logically through the alphabet has, in fact, been highly successful. Braille is still the most widely used tactual reading system throughout the world and is preferred by competent, fast blind readers. But braille characters are usually recognised faster by moving the forefinger gently across the dots. That produces a lateral shear pattern on the ball of the finger (see later).

The general assumption that braille characters are initially perceived most easily as holistic shapes by active touch was tested experimentally by using a number of widely different but converging methods. None of the findings supported the assumption. A brief summary is given below.

In vision, identical shapes are matched faster and more accurately than different shapes, either because the same global form makes identical shapes easier to recognise (e.g. Krueger, 1973), or because identical shapes prime each other (Nickerson, 1965). That was not the case for braille characters in touch. Sighted children who had never seen or felt braille letters before could discriminate four relatively distinct braille letters quite accurately by touch without vision. But they were no faster or more accurate in judging identical than different braille shapes by touch. In vision precisely the same task and stimuli produced the usual advantage for identical shapes (Millar, 1977b). Even after having learned to name four tactually explored braille letters, sighted children had little idea of their shapes when drawing the letters from memory (Millar, 1985b). As noted earlier, single braille patterns produce quite easy visual shapes.

The finding that children confuse braille patterns that differ only by one dot, or by the position of one dot, is also sometimes considered evidence that they code their outline shapes (Nolan & Kederis, 1969). However, errors due to missing dots, or failure to assign dots to their correct location, indicate poor discrimination early in learning, rather than early contour perception. When tested directly with raised outline shapes, beginning young braille readers recognised the letters more easily in dot format than as raised outline shapes (Millar, 1985b). They also judged the "odd one out" of three words much more easily by the sound or the meaning or the dot density of the words than by their outline shape (Millar, 1984a).

Other features that are typically easy in vision, such as the size of patterns, the relative location of shape features, and symmetry, provided further tests. Beginning young braille readers were significantly less accurate in recognising enlarged braille letters than in recognising the same letters in their normal small format (Millar, 1977a). More experienced readers were less disadvantaged than beginners in recognising letters in enlarged format. But even they were faster with the more familiar, conventional format. In visual shape perception, the location of distinct features relative to each other is obvious. In braille pattern recognition by touch, differences in the number and spacing or density of dots in a pattern, rather than in their spatial location, size or shape, improve recognition (Millar, 1978, 1985b, 1986).

Symmetry greatly facilitates visual shape recognition. In touch, however, the ease or difficulty of a dot pattern depends on the density of dots in patterns, and not on symmetry (Millar, 1978). Interestingly enough, blindfolded participants detected haptic shape symmetry only when explicitly instructed to relate the starting position and scanning movements of the two fingers to their body midline (Ballesteros, Millar, & Reales, 1998). Even so, the difference was confined to open shapes, suggesting that shape features were also a factor.

There are other differences in detecting shape features by touch than by vision also with somewhat larger figures than braille (for a review see Gentaz & Hatwell, 2000). Some of these seem to be due to relating a given feature only to the hand rather than to more balanced reference cues.

Let me emphasise that these differences do not mean that braille patterns "cannot" be coded as shapes by touch, or that practice with a small set of larger patterns does not make it easier to recognise the identical smaller patterns subsequently. Neither blindness, haptic perception nor the physical features of braille make children incapable of generalising stimulus information. This emphasis is needed because findings with braille (e.g. Millar, 1997) are often misinterpreted as if they were assertions of inability or were meant to imply that vision and touch have nothing in common.

However, findings showing that finger movements do not merely get faster with practice, but actually change in the features that are picked up by experienced braillists, are important for understanding underlying cerebral function (see later). Fast braillists depend on dynamic lateral scans in reading, and they use quite different movements for letter-detection tasks. The evidence was obtained by video-recoding the hand and finger movements from underneath transparent surfaces and is reviewed in the next section.

SPATIAL CODING AND LATERAL SCANNING MOVEMENTS IN TEXT READING

The pick-up of perceptual information by touch and movement is highly relevant to the question of how the spatial aspects of small patterns are coded and how that is best tested.

Data from an on-line video recording and timing device showed the importance of different types of hand and finger movements relative to the orientation of the text and to the orientation of the body in the pick-up of haptic information with different braille reading tasks (Millar, 1987a, 1988b, 1997). The device (Figure 4.2) films hand and finger movements on braille texts from underneath transparent surfaces. An integrated digital video-timer simultaneously records precise (1/100 s) cumulative time. A microphone is attached, and auditory inputs are recorded concurrently. A video monitor shows the finger movements above the text and the concomitant cumulative real time above the text. The larger of two connecting units contains the

Figure 4.2 Photograph of the recording device, seen from the participant's side. The experimenter sits on the left, facing the monitor, video-timer and recorder. The participant sits in front of the open side and reads the raised braille text reflected into the inverted (scan coils reversed) camera. [From S. Millar (1997), *Reading by Touch*. Published by Routledge (Taylor & Francis), London. Reproduced with permission.]

78   SPACE AND SENSE

plate-glass reading surface, facing the participant. It is fixed at normal table height above an angled mirror that is flanked by two strip lights.

Texts are brailled onto sheets of transparent drawing film that produce raised dots that are darkened from the reverse side. These are fastened to the reading surface. The smaller unit houses the video camera facing the mirror. The scan coils of the camera are reversed and the camera is inverted, so as to produce normally oriented pictures on the monitor. The movements of each finger on, and also above, the text are clearly visible on the monitor. The precise cumulative real (1/100 s) time is displayed above the text in each frame. The midpoint of the white patch that appears on the ball of the finger that touches a braille dot or a gap between dots is seen on the monitor. It provides the reference location for the precise time in the (40-ms) frame-by-frame analyses in replay (Figure 4.3). The frame time, above the text, is transcribed onto braille text sheets for every dot touched, separately for each hand (see later, Figures 4.6a and b). People can thus read braille texts normally with either or both hands, while precise data on the time and the location of any dot on each finger are recorded.

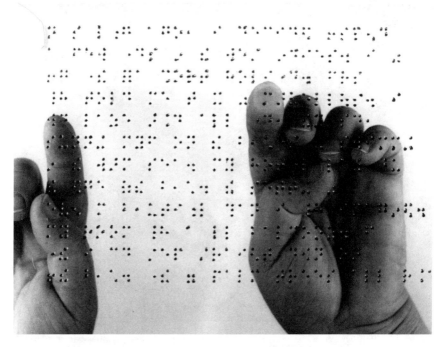

Figure 4.3   Still photograph of the braille text and reading hands. The right forefinger is on the braille letter "m" on the text and shows the typical white patch of the ball of the finger when touching. [From S. Millar (1997), *Reading by Touch*. Published by Routledge (Taylor & Francis), London. Reproduced with permission.]

The time spent on any one character, or even on a dot within a character, and movement time can be calculated accurately from these protocols.

The device produced important findings about the type of perceptual features that are picked by touch and movement and how they vary with the reading task and proficiency. Grunewald (1966) had suggested that fluent readers use dynamic lateral scanning. Such scanning produces lateral shear patterns on the ball of the reading finger. The hypothesis was tested in a double-dissociation design against the traditional view that braille is read letter-by-letter based on the shape of each letter and that it simply gets faster with proficiency.

Shear patterns across the ball of the reading finger were tested by changing the orientation of the finger to the text in reading the same (counterbalanced) texts for comprehension and in letter-search tasks (Millar, 1987a). The lateral scan hypothesis predicts that disrupting the lateral shear pattern on the ball of the reading finger would affect fluent readers most, particularly in reading texts for meaning, but less in letter search in the same texts. On the other hand, if braille depends on letter-by-letter shape perception at all levels of proficiency, and simply gets faster with experience, disrupting shear patterns should not affect prose reading more than letter search at any level of proficiency. All experimental texts were rotated by 90 degrees relative to the reader's body midline. The rotated texts required near-to-far instead of left-to-right reading. The unfamiliar orientation was, of course, expected to slow reading in all conditions. But the orientation of the finger to the text was crucial. The arm, hand posture and finger posture had to be inclined, somewhat awkwardly, to preserve the lateral scan from near to far. Keeping the finger in the normal near-upright position relative to the body and to the text disrupts the shear pattern on the finger, but it would make less difference for quasi-circular movements over letter shapes. The results for three proficiency groups are shown in Figure 4.4, in which all real-time latencies have been converted to word-per-minute (wpm) rates for comparison.

The findings provided the first experimental evidence that haptic perception does not merely become faster, but that different features are picked up with proficiency. Moreover, the features that are picked up by proficient braille readers depend on the type of reading task being used. But it would differ less, or not at all, for slow readers. Grunewald's (1966) hypothesis was well supported. Fluent readers used lateral scanning in reading for meaning, but not for letter search. Slow readers depended on movements that produced letter-by-letter reading. The pick-up of stimulus features in haptic perception thus differs with proficiency and types of reading task.

For slow reading there are almost more hypotheses about the role of the two hands than about styles of reading. In fact, none of them is wholly wrong. Reading styles differ with speed, proficiency and previous reading experience (see later) as well as with the reading task. It has been suggested

80 SPACE AND SENSE

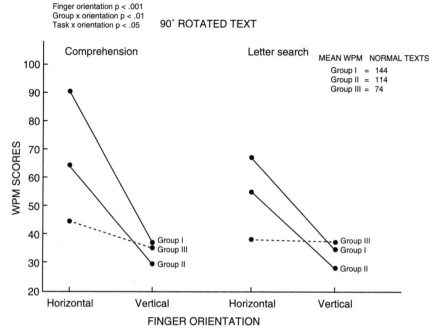

Figure 4.4 Mean wpm rates in reading for comprehension and letter search tasks in lateral scan (finger inclined horizontally) and vertical (finger positioned normally) relative to the rotated text. The mean wpm rates for the three proficiency groups are shown in the top right-hand corner. [From S. Millar (1997), *Reading by Touch*. Published by Routledge (Taylor & Francis), London. Reproduced with permission.]

that keeping the fingers together on a line in reading provides "a wider window" or perceptual view (Bürklen, 1932; Kusajima, 1970). That is doubtful for young readers who are slow or retarded despite having learned braille from the start, because they tend to rub unsystematically with many regressions over characters when they fail to recognise words or characters (Millar, 1997).

The finger movements of sighted people who have learned braille as adults are noteworthy. Adults who used to read print fluently tend to be slow, letter-by-letter braille readers, although they can cope with semantically difficult texts. Video-recordings showed that they do indeed explore the shape of each character when reading. Circular movements, down/up strokes or diagonal motions over each letter are common. It seems likely that late blind readers expect or imagine the letters to form distinct visual shapes, at least initially. Interestingly, it was a former print reader who had advocated passive touch as a means of perceiving braille characters as shapes. In principle, spatial coding is possible by resting the finger lightly on a pattern. The top and sides of the fingertip could act as a literal reference frame for locating the dots in the pattern in letter-by-letter reading. Most adult former print readers prefer to begin braille with enlarged braille characters (Tobin, 1988), possibly because the patterns require somewhat larger top-down and circular scanning movements that make it easier to construct shapes that can be envisaged. The habitual scanning movements can then be transferred gradually to conventionally sized braille.

Differences between early and late braille readers are likely to be due to long-term reading habits that former fluent visual print readers bring to learning braille. Some of these probably interfere with the development of the most efficient hand movement in learning to read braille late in life, even if there has been a good deal of experience with touch. Early experience with braille still results in better discrimination by older people, despite the process of aging (Stevens, Foulke, & Patterson, 1996).

The role of lateral scanning and of the different movements that are used to find a single character is important in understanding the relation between touch and vision. The effect may relate to studies that have explored brain activities with non-invasive techniques. These have shown that touching braille patterns activates visual areas in the occipital cortex of early blind braille readers more than in late blind readers or in sighted people (Büchel, Price, Frackowiak, & Friston, 1998; Burton et al., 2002; Sadato et al., 1996).

However, tactually sensitive cells in regions within the areas in the ventral stream in visual regions in the occipital cortex are also activated in tactual object- and shape-recognition in the sighted (e.g. Amedi, Malach, Hendler, Peled, & Zohary, 2001).

The main assumption has been that the effects are due to crossmodal plasticity and a take-over of function by the cells in the absence of sight. A possible explanation is that the population of cells within the visual cortex that are activated by tactual stimulation expands so that adjacent formerly vision-sensitive cells become activated by touch, similarly to the expansion found in the sensorimotor cortex in representing the reading finger in braille (Pascual-Leone & Torres, 1993). Evidently, quite transient experience can make a difference even in adulthood. A transcranial magnetic stimulation (TMS) study, for instance, showed rapid modulation of motor cortical outputs that depended on hours of proofreading compared to resting without on the preceding day by proficient braille readers (Pascual-Leone, Wassermann, Sadato, & Hallett, 1995). However, the factors that actually produce either transient or long-lasting or possibly permanent changes may be more complex than that.

Sadato and colleagues (1998) explored the neural networks that are used for braille reading. They measured regional cerebral blood flow in positron emission tomography (PET) studies. Proficient early blind braille readers discriminated uncontracted braille words versus braille letter-strings. The studies showed activation in a network that included the ventral occipital regions and primary visual cortex. But there was no activation in the secondary somatosensory area that is normally involved. The effect was evidently irrespective of which (right or left) was the reading finger.

A number of studies have reported enhanced tactile acuity, or "sharper" tactile perception, by early blind people (van Boven, Hamilton, Kauffman, Keenan, & Pascual-Leone, 2000; Grant, Thiagarajah, & Sathian, 2000; Goldreich & Kanics, 2003; Sathian & Zangalade, 1996), although the methods of assessing acuity have varied.

In view of the findings on the pick-up of braille features by early blind fluent readers, the neurological findings raise important further questions that are considered briefly in the final chapter.

A further question arises also from findings that showed that fluent braille readers use both hands alternately for the verbal and spatial aspects of text reading. The smooth, rather elegant, lateral scanning movements of fluent braillists in reading prose texts have impressed many observers (e.g. Kusajima, 1970). Seen from above, it looks as if the fluent readers use the two hands simultaneously, often on two different lines of text. One view was that reading is fast because the two hands read two different lines of text at the same time (Bertelson, Mousty, & D'Alimonte, 1985; Bürklen, 1932; Kusajima, 1970). The alternative view was that the two hands perform different functions alternately (Foulke, 1982).

Evidence on the respective functions of the two hands was obtained from the detailed time relations between the left and right forefingers in two-handed prose reading from (40-ms) frame-by-frame analyses of video recordings. The minimum criterion for simultaneous reading of different verbal information on successive lines of text was that each forefinger moves to at least one new letter simultaneously on two different lines of text. The frame times for transitions times for the two fingers to move from one line to the next in reading prose texts were used with competent young (mean age 17 years) braille readers (Millar, 1987b). If the two hands do read different texts simultaneously on the two lines, the left finger should move to at least one new letter on the next line before the right hand has finished reading the previous line.

As will be seen from Figure 4.5, the reverse time relation was found. For each of the 10 participants, the right finger either touched the last letter on the previous line, or the blank margin beyond the text, or was in transit above the text to join the left finger, some hundreds of milliseconds before the left finger had moved from holding on to the first letter on the next line to even one new letter.

The findings showed that fluent braillists do not use the right and left hand to read different texts simultaneously on two lines of text. They do not even pick up information from two new letters at the same time. The hands

4. HAND MOVEMENTS AND SPATIAL REFERENCE 83

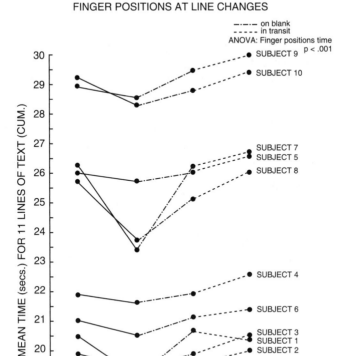

Figure 4.5 Finger positions at line changes: The left has found the start of the next line (Left NL NL); the right finger is on the last letter of the previous line (Right, last letter, last line); the right finger moves off the last letter (Right OFF); the right is in transit and then joins the left hand (Right NL) on the new line. [From S. Millar (1997), *Reading by Touch*. Published by Routledge (Taylor & Francis), London. Reproduced with permission.]

perform different functions simultaneously. Division of labour for the verbal and spatial aspects of reading, rather than simultaneous intake of verbal information, characterised fluent reading and was also shown by other analyses of braille protocols (Millar, 1997). Figures 4.6a and 4.6b are transcript examples of fluent text reading. The right hand reads the right half of a line of text after the left hand has left to locate and hold on to the start of the next line. The left hand starts reading the new line after the right has finished reading the previous line. It moves off from the previous last line and makes a fast transition move, above the text, to join the left hand on the

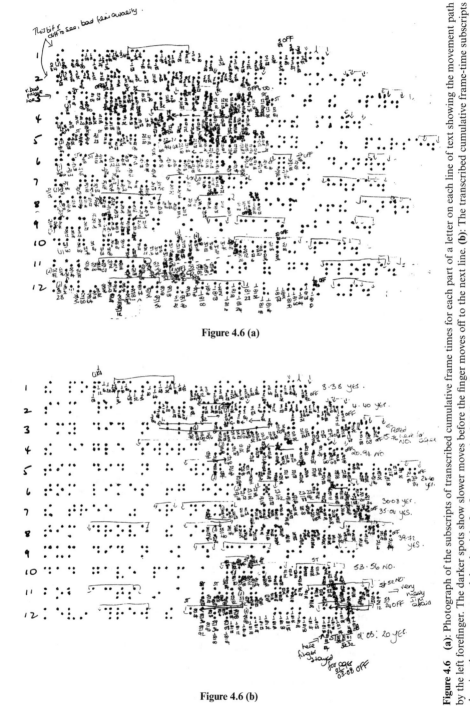

Figure 4.6 (a): Photograph of the subscripts of transcribed cumulative frame times for each part of a letter on each line of text showing the movement path by the left forefinger. The darker spots show slower moves before the finger moves off to the next line. (b): The transcribed cumulative frame-time subscripts showing the movement path by the right forefinger on each of 12 lines of text.

new line. The changeover occurs roughly in the middle and at the end of each line of text. It is too fast to be observed by looking at the hand movements from above. The protocol transcripts of frame times onto braille sheets show the portions of text typically read by each hand in fluent text reading (Figure 4.6). The frame-time transcripts show that the left hand reads the left portion of each line of text and the right hand reads the right portion of each line of the text.

Typically, while the left hand is reading the new line, the right hand is engaged in moving from the previous line to join the left hand on the next line. The right hand takes over from the left hand after having joined it on the new line. The left hand then moves to the start of the next line. The left hand typically feels for the start of the next line and holds on to it, while the right hand reads the rest of the previous line. The left hand starts reading a new line again after the right hand has reached the last letter of the previous line, or is briefly feeling the blank beyond the text, before moving off again, usually above the text, to join the left hand again on the next new line of text.

Both hands thus alternately read to take in verbal information while the other hand is engaged in spatial tasks such as finding the next line or joining the other hand. The division of labour is a hallmark of fluent reading.

The hand movements of slow readers differ in many respects. Slow readers who use both hands tend to move the two forefingers side by side on each line of text. But only one hand is used for reading. In left-handed reading, the right is in advance and guides the reading finger along the line of text, or keeps the place during the many regressions that are common in slow reading. In right-handed reading, the left hand usually remains stationary at the beginning of the line of text and then moves down one line when necessary. Letter-by-letter reading is common. Regressions are symptomatic of reading with non-optimal finger movements.

By contrast, competent readers occasionally use several fingers together. If they do, it is usually to gauge the length of a long word when they are unsure of its meaning, or because the braille dots have become faint with use or have been depressed deliberately in experiments that test such conditions (Millar, 1997).

Longer-term experience with haptic materials compared to visual experience also affects perception of their orientation (Heller, Calcaterra, Green, & Barnette, 1999; Heller, Calcaterra, Green, & de Lima, 1999). Long-established habits are difficult to alter. But it is as yet an open question whether it is impossible. Equating control groups in terms of years of experience can be misleading. One of the best formerly sighted print readers whom I knew had learned braille for about eleven years and had developed a zigzag motion, in scanning lines of text, that seemed to pick up both shape and shear patterns.

The fluent young readers who had learned braille from the start were ten times faster, although they had also started braille only about twelve or thirteen years earlier. Nevertheless, their experience of braille was vastly greater. Children whose first written language was braille had used it every school-day, and for every subject throughout their childhood and adolescence.

Taken together, the braille studies show that shape features that are most easily perceived in visual print shapes are not necessarily the easiest in haptic perception. At the same time, visual shape features "pop-out" pre-attentively in perception, without any need to search, only in some conditions (Treisman, 2004). The differences between visual print and braille must be attributed, at least partly, to the physical composition of the pattern that was chosen by Louis Braille. The reason why it is still the preferred system by the blind shows the importance of finger and hand movements in haptic perception. The adaptation of finger movement to serve the different functions that reading tasks require is shown by fluent readers

The important point of the findings on the scanning movements is that they document some of the details in which the same activity – namely reading – differs between active touch and vision. But the findings also show that touch, in conjunction with movements of the two hands, can achieve almost the same facility and speed as visual reading. At the same time, the change in the use of the two hands also related significantly and differently to the linguistic and spatial demands of the reading task. It can be assumed, therefore, that the change involved reciprocal interactions between linguistic/ cognitive and perceptual factors (Millar, 1990).

The behavioural findings thus suggested a complex network of reciprocal relations between the physical features of the braille system and the finger movements that maximise inputs from touch for the diverse linguistic, movement and spatial demands of different reading tasks, which, in turn, affect the pick-up of perceptual information,

The change in scanning movements with proficiency also related significantly, and differently, to the spatial demands of the reading task. The alternate verbal and spatial roles assigned to the left and right hands in text reading implies changes in, or, at least, the dual involvement of, the right and left cerebral hemispheres with proficiency.

It raised the next question – namely how far it is possible to infer spatial coding purely from performance by the two hands.

WHICH HAND IS BEST? SPATIAL CODING WITH THE LEFT AND RIGHT HAND

There is ample evidence that spatial information is processed predominantly in circuits in the right cerebral hemisphere and subcortical regions, whereas circuits in the left cerebral hemisphere are dominant for processing linguistic information, at least in right-handed people (e.g. Gazzaniga, 1988; Gazzaniga & LeDoux, 1978; Kimura, 1973; Kimura & Dunford, 1974).

Damage to circuits that include parietal regions in the right cerebral hemisphere impairs visual recognition of the right half of space. There have been fewer studies with touch, but damage to the right hemisphere also impairs tactile recognition on the left side of the body. Damage to similar circuits in the left half of the brain leads to neglect of the right side of space. But effects tend to be less severe or do not last as long, possibly because ipsilateral fibres come to be recruited (Behrmann, 2000; Berthoz, 1991; Bisiach & Vallar, 1988; Paillard, 1991; Rains & Milner, 1994). The majority of nerve fibres, including fibres from somatosensory sources that originate on one side of the body, cross over to the opposite cerebral hemisphere (e.g. Rains & Milner, 1994). In principle, therefore, the right side of the body, including the hand, should be better for language tasks, and the left side, including the left hand, should be better for shape and spatial coding. But findings vary.

It is fair to say that by far the largest number of studies have used visual information in conjunction with movements as outputs. But there is general agreement that the right-hemisphere dominance for processing spatial information from the opposite side of the body also applies to touch, although the left hemisphere can take over when necessary (Milner & Taylor, 1972; Smith, Chu, & Edmonston, 1977). Better recognition of wire shapes with the left hand has been reported after cerebral comissurotomy (Kumar, 1977).

The notion that the left hand is superior for the pick-up of shape and spatial information by touch is widespread among teachers of braille. It is a confident assumption, even by some serious researchers, that one can infer shape and/or spatial processing directly from any left-hand advantage by right-handed people. Indeed, the left hand has actually been called "the eye of the blind", on the ground that the right hemisphere is more involved in spatial processing than is the left.

Nevertheless, empirical findings on hand effects in studies of braille have been highly inconsistent. Some found the left hand better for naming single letters (Hermelin & O'Connor, 1971; Rudel, Denckla, & Hirsch, 1977). Others showed that the right hand of experienced braillists is superior for reading (Fertsch, 1947) or found no advantage for either hand early in learning (Bradshaw & Nettleton, 1981; Bradshaw, Nettleton, & Spehr, 1982; Millar, 1977a). Some findings have suggested that the left hand may be better for discrimination and the right hand for letter naming (Millar, 1984b; Semenza, Zoppello, Gidiuli, & Borgo, 1996). Since the left hemisphere (Broca and Wernicke's areas) is dominant for phonological, lexical and semantic language skills that are essential in learning and reading braille, discrepancies in findings for different braille tasks are perhaps not surprising.

For braille, the relations between the hand being used for the pick-up of perceptual information and the spatial, verbal and movement processes that are involved in different types of reading tasks and at different stages of learning and proficiency are too complex to permit simple inferences about functions from performance by either the right or left hand.

However, it seemed likely that the discrepancies in findings on braille are mainly due to the combination of the physical characteristics of braille, linguistic factors in learning to read and effects of exploratory and scanning movements.

Hand effects in haptic perception of larger, non-verbal displays were therefore important, and they are considered next.

HAND MOVEMENTS AND SPATIAL REFERENCE FOR LARGER NON-VERBAL DISPLAYS

The question was whether performance by the left hand constitutes a sufficient behavioural test of haptic spatial coding with non-verbal tasks. Given the general finding that active touch is superior to passive touch (e.g. Gibson, 1962; Heller & Myers, 1983; Katz, 1925), it seems reasonable to briefly consider behavioural findings on active touch, which involves finger and hand movements, compared to those on passive touch, which does not.

Movements are mainly processed in the left cerebral hemisphere (De Renzi, Faglioni, Lodesani, & Vecchi, 1983; Kimura, 1993). A clinical study showed a left-hand/right-hemisphere advantage for remembering locations from touch with passively guided exploring movements (De Renzi, Faglioni, & Scotti, 1969). Right hemisphere posterior parietal and inferior intraparietal sulcus regions dominate in visuo-spatial orienting responses. But movements involve the left hemisphere (De Renzi, 1978, 1982). Studies of brain lesions, and non-invasive fMRI (functional magnetic imaging), PET (positron emission tomography), and TMS (transcranial magnetic stimulation) techniques have shown that attention to movements involves the posterior parietal cortex and related circuitry more in the left than the right cerebral hemisphere (Rushworth, Nixon, Renowden, Wade, & Passingham, 1997). Motor attention and imagery of hand movements engage left hemisphere processing in regions that are somewhat anterior in parietal regions (Rushworth, Ellison, & Walsh, 2001; Rushworth et al., 1997; Sabate, Gonzales, & Roderiguez, 2004). A PET study suggests that left hemisphere prefrontal, premotor and parietal regions are dominant for action generally, irrespective of the hand used (Schluter, Krams, Rushworth, & Passingham, 2001).

Such findings might suggest that the right hand performs better in active touch, and that the left hand is better for passive touch. There was some behavioural evidence for that. Lechelt (1982) showed that the left hand was better for passive identification of the location of dots in a matrix, but that the right hand was superior in precisely the same spatial task when active scanning was used. Similar results were reported in another study (Smith, Chu, & Edmonston, 1977). Alternatively, it seemed possible that superior performance by the left hand may be due to greater tactile sensitivity to touch stimuli by the left hand than by the right hand. Touch thresholds, tested by judging being passively touched by one or two stimuli, were lower for the left hand (Delantonio & Riggio, 1981; Riggio, Dellantonio, & Barbato, 1980). But the findings on hand effects in passive versus active touch do not provide a univocal behavioural test of haptic spatial coding.

There was also other evidence that even tests with obviously "spatial" tasks do not necessarily constitute a sufficient behavioural criterion for inferring purely spatial processing. Standard tests of spatial ability, such as mental spatial rotation, spatial inference and spatial reasoning tasks, also require logical inference and reasoning. Performance on such standard tests of spatial ability is impaired by lesions in the left cerebral hemisphere as well as in the right cerebral hemisphere in visual conditions (Langdon & Warrington, 2000; Serrati et al., 2000).

Even apparently simple matching tasks suggest some left hemisphere involvement (Mehta & Newcombe, 1991). Hemi-neglect of the left side of space involves higher-order cognitive functions, but could also include more peripheral cross-lateral effects (Bisiach & Vallar, 1988). Indeed, spatial processes are often difficult to distinguish from other cross-lateral effects (Stein, 1991).

Behavioural investigations that rely on interpreting a left-hand advantage as evidence for spatial coding thus have to consider what factors, other than purely tactile inputs, may be involved in given spatial tasks. For instance, haptic shape and spatial tasks that involve sequential versus simultaneous inputs could involve the two hemispheres unequally (Minami, Hay, Bryden, & Free, 1994; Nicolls & Lindell, 2000).

The apparent discrepancies in findings that relate better processing with the left or right hand to spatial coding suggested that a behavioural test of coding in active touch was needed that would test spatial coding and hand effects independently.

HAPTIC SPATIAL CODING AND HAND EFFECTS: A TEST OF TWO HYPOTHESES

Exploratory hand and finger movements have to be systematic for accurate haptic perception, and that can evidently be achieved by instruction and practice in specific tasks (Berlá & Butterfield, 1977). Unsystematic exploration early in learning (Vurpillot, 1976) and/or with unfamiliar material is not confined to small braille patterns. Berlá and Butterfield (1977) found that blind students failed to recognise unfamiliar geographic shapes from raised contours in tactile maps. They were able to do so once they had been trained to explore the contours systematically.

The importance of reference frames for accurate coding of movements was stressed by Jeannerod (1988), Paillard (1991) and Berthoz (1993). Similarly, the criterion of spatial coding here turned on the use of reference cues (Chapter 2; see also Behrmann, 2000; Millar, 1994; Paillard, 1991). Systematic exploration requires at least one anchor or reference point that can be recognised as the end and starting point of the exploratory movement. For hand or finger movements to become systematic and accurate, the movements have to be coded spatially by reference to start and end positions.

Operational criteria for what is to count specifically as "spatial" processing, independently of hand effects and task factors, should be potentially useful for clinical assessments and rehabilitation. These criteria are needed to understand how hemisphere asymmetries in spatial coding relate to other cross-lateral hand and task effects.

The study described here tested two hypotheses. The hypothesis that the left hand is better for spatial tasks predicts a left-hand advantage in all conditions. By contrast, the reference hypothesis predicts significantly greater accuracy in haptic recall with explicit additional reference information than in conditions that do not provide explicit reference information. It makes no *a priori* predictions about hand advantages.

In order to test additionally how scanning movements relate to reference information, it was decided to use tasks of judging distances and tasks of judging locations. The point was to use the location task, rather than the distance task, to test for movement effects. Long, complex and discrepant positioning movements are known to disturb the recall of locations (e.g. Laabs & Simmons, 1981; Millar, 1985b; Millar & Al-Attar, 2001; Wallace, 1977). Moreover, the reference hypothesis assumes that distance judgements, as well as location judgements, are spatial tasks. That may seem to be at variance with the notion that haptic distance judgements depend solely on kinaesthetic inputs. However, movement distances should be coded spatially if they can be related to reference information. A previous study (Millar & Al-Attar, 2003a) had shown that haptic distance judgements do involve spatial coding. Recall of a repeated small distance was significantly disturbed, not only by an irrelevant movement task, but very significantly also by a spatial task that required no movements. The location, rather than the distance, task was therefore used here to test for movement effects. A longer, more complex positioning movement was needed for location recall than in presenting them initially. For the distance task, a repeated small positioning movement was used instead in presentation and in recall. In all other respects, the experimental conditions were the same in the two tasks (Millar & Al-Attar, 2003b).

The control condition in both tasks consisted of scanning the critical distances, or the critical locations, in presentation and recall without touching any other part of the display or surrounds. In the experimental or reference conditions in both tasks, participants were instructed to use an actual external frame around the stimuli, and also their body midline for reference.

Participants were 40 right-handed high school volunteers who were tested blindfolded and allocated randomly to different conditions. To avoid alerting participants to the use of reference cues in control conditions, and at the same time equating conditions in the tasks of judging distances and locations as much as possible, half the participants used either the right or the left hand for scanning in both tasks, being tested either first on distances or first on locations. The same procedures were used with participants who received reference-frame instructions. The experimenter always placed the participant's scanning finger at the start of scanning.

The usual method of testing distance recall is to start the recall movement from a different location. The point of the displacement is to make it more difficult to code distances in terms of start and end locations. The distance task required recall of two short embossed linear distances in a raised-line map-like tactual layout (Figure 4.7). One distance was scanned from left to right, the other from right to left.

In control trials participants scanned only the relevant distances. The frame surrounding the map-like display was never touched, and there was no mention of reference cues at any time. Precisely the same procedures were used in experimental conditions that provided reference information. But participants were shown in pre-tests how to relate the to-be-remembered distance to their body midline. They were also shown how to use their free hand to feel the rectangular frame that surrounded the map. The instructions stressed that both should be used as means of remembering the critical distance.

In recall tests participants were to stop the scanning finger when the remembered distance was reached, in both control and reference conditions. The experimenter used replica maps with ruler markings that the blindfolded participants could not see. The stopping point of the finger in recall was marked on the replica maps for each of four left–right and right–left distance trials, respectively, in control and reference conditions. Absolute errors were calculated from the correct distance to participants' stopping points. The results for the distance tasks are graphed in Figure 4.8.

The graph shows clearly that the added reference information reduced errors very significantly compared to the control conditions, regardless of whether the left hand scanned the distance in control and frame conditions and the right hand was used for the frame, or whether the right hand scanned

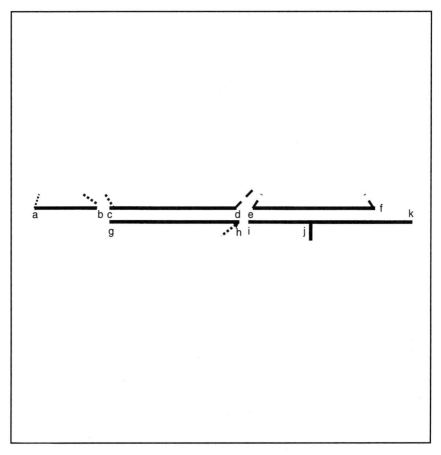

Figure 4.7 Test map for the 3-cm distance task (no letters in actual map): presentation a–b starts at c in recall; presentation j–i starts at h in recall, The surrounding frame was only used in the reference instruction condition. [From S. Millar & Z. Al-Attar (2003), "Spatial reference and scanning with the left and right hand", *Perception*, *32*, 1499–1511. Reproduced with permission from the publisher, Pion Ltd., London.]

the distance in control and reference conditions and the left hand was used for the frame. The left and right hands did not differ significantly from each other in accuracy in either control conditions or in reference instruction conditions.

The results clearly supported the hypothesis that the use of external-frame and body-centred reference cues makes haptic distance judgements more accurate. The fact that the accuracy of recall with the left hand did not interact differentially with the significant increase in accuracy with the instructions to use reference cues was not, perhaps, surprising. Testing recall of the distances depended on scanning movements and on relocating that distance

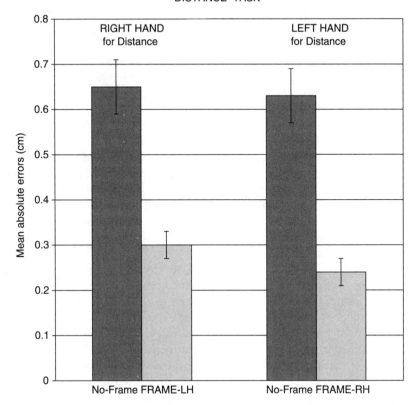

Figure 4.8 Mean errors and standard errors in the distance tasks for the No-frame (no reference instructions) and the Frame (reference instruction) conditions for scanning with the right or left hands. [From S. Millar & Z. Al-Attar (2003), "Spatial reference and scanning with the left and right hand", *Perception*, *32*, 1499–1511. Reproduced with permission from the publisher, Pion Ltd., London.]

information. Scanning movements to reproduce the distance should be the same in recall as in presentation. Scanning the distance would thus involve left-hemisphere processing of the movements, as well as the spatial aspects of relocating the end position from the new (guided) starting point, and therefore right hemisphere processes also. Cross-lateral effects from both right and left hemisphere processes that inhibit or counterbalance each other would explain why the left hand did not perform better than the right and why it did not relate differentially to the advantage in accuracy from instructions to use spatial reference cues.

As noted earlier, the usual method of testing the recall of distance information is to start the recall movement from a different location. The point of the displacement is to make it more difficult to code distances in terms of start and end locations. The significant increase in accuracy in experimental conditions suggested that the instructions to use external and body-centred reference cues made it easier for participants to relocate the start and end positions of the distance relative to body-centred and/or frame cues.

The location task was used specifically to test for movement effects in addition, to see how these would relate to the reference conditions. Simple, short positioning movements were therefore used to reach the to-be-remembered locations in presentation. Longer, directionally reversed and more complex movements that included a gap in the raised route were required to reach the remembered criterion locations in the recall phase.

It was expected that the more complex movements would make location recall more difficult than the distance task overall. Given that movements are processed more readily by the left hemisphere than by the right hemisphere, right-hand scanning of locations should be less disturbed by complex positioning movements in recall than scanning with the left hand, although that difference might also be diminished by a right-hemisphere advantage for spatial coding for the left hand.

The same map-like display that was used for distance judgements was used for locations, with the sole addition of two embossed shapes on each of the raised-line routes. The middle of each shape was the to-be-remembered location. The shapes were not present in the recall map. One location required a short left–right movement in presentation, but it could only be reached in recall by a longer left–right movement that included a small gap. The other location was reached by a short right–left movement in presentation, but it could only be reached in recall by a longer right–left movement that included a short gap. The other procedures were precisely the same as for the distance task. Figure 4.9 shows the results for control conditions, which differed only in providing explicit instructions to use the external-surround and body-centred cues as reference information for the target locations.

The differences in recall accuracy between control conditions and conditions that provided explicit reference information were again highly significant. The differences were as large for scanning the locations with the right hand and using the left hand for the frame, as for using the left hand for scanning the locations and the right hand for the frame. There was a hand effect. As expected from the longer and more complex scanning movement in recall, the recall position was more accurately shown with the right hand than with the left. But the difference in errors between the hands reached significance only in the two reference conditions. More important, the hand effect did not interact with the highly significant increase in accuracy with the reference conditions.

The subsequent overall analysis of distance and locations tasks showed that errors in the location task were significantly larger than in the distance task, as had been expected from making scanning in recall more difficult in the location task than in the distance task. The overall larger error and

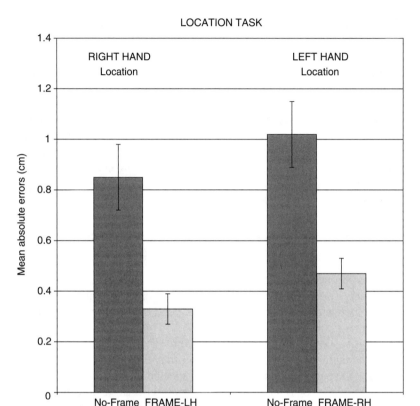

Figure 4.9 Mean and standard errors for the location tasks in the No-frame (no reference instructions) and the Frame (reference instruction) conditions for scanning with the right or left hands. [From S. Millar & Z. Al-Attar (2003), "Spatial reference and scanning with the left and right hand", *Perception*, *32*, 1499–1511. Reproduced with permission from the publisher, Pion Ltd., London.]

right-hand advantage in the location task could safely be attributed to the (deliberate) discrepancy in the recall route, because the same participants had taken part (in counterbalanced fashion) in both tasks.

The overall analysis confirmed the highly significant effect of the reference condition in both tasks and the fact that it did not interact significantly with effects of the left hand versus the right hand. The lack of interaction of hand effects with reference conditions also showed that the effects of the reference conditions were independent of the (modest) right-hand advantage with discrepant recall movements.

The important finding was thus that instructions to use body-centred and external-frame cues for reference improved recall accuracy very decisively in both tasks, independently of hand performance, task differences and movement effects.

The increase in the accuracy of recall for both the distance tasks and the location tasks with the use of reference cues provides empirical support for the validity of the operational criteria for spatial coding that the reference hypothesis implies. The finding also suggests that explicit reference information can be used as a reliable behavioural test of spatial coding (Millar & Al-Attar, 2003b).

As far as I know, cross-lateral hand effects had not previously been tested with independent behavioural spatial criteria in studies of haptic performance. The results show that the criteria used in the study discussed here can provide a useful behavioural method for assessing spatial processing in studies designed to investigate these and other cross-lateral effects further, including different forms of sensory registration.

SUMMARY: SPATIAL CODING AND MOVEMENT INFORMATION IN HAPTIC PERCEPTION

If I have succeeded in describing the functions of finger and hand movements in haptic perception at all accurately, albeit not exhaustively, it should be clear that, combined with touch, the functions depend on a rather large number of interdependent factors.

The physical composition of the material itself, the form of discrimination to which the material lends itself and how it may be coded, the size of inputs, the degree of experience by individuals of specific stimulus information, and previous experience of what is (e.g. linguistically) the same task, but differs in medium and modality, are some of the factors that determined outcomes in braille letter detection versus reading for meaning.

The different types of scanning movements that were found for different aspects of a task are a further important factor. The findings showed that different scanning movements pick up quite different perceptual features in different tasks. The stimulus input in lateral scanning evidently depended on temporal/sequential features. But the lateral finger movements also differed in spatial direction from the up/down and circular movements over single characters found in letter search tasks. The change in the braille features that are perceived with experience in different tasks suggests that positive and negative feedback from exploring movements in picking up expected stimulus information plays some part in diversifying finger movements. "Top-down" effects as well as the "bottom-up" pick-up of stimulus features must thus be assumed to influence their deployment in turn (Chapman, Tobin, Tooze, & Moss, 1989). The findings have practical implications for learning braille. Teachers of braille should be alerted to differences in the pick-up of tactual features with different scanning movements. The information could be made available to the learner by encouraging lateral scanning of words or sentences, or by direct instruction.

"Top-down" effects must also be assumed in the deployment of the two hands relative to each other. Using the two hands alternately for the pick-up of verbal versus spatial information was typical of fast text reading. As noted earlier, it is doubtful that really fluent and fast readers are any longer aware what their fingers are doing in reading for meaning, or of their finger movements in searching for a particular letter in a text, or how they deploy the two hands, unless they are asked to reflect on how they read. It may be taken as a hallmark of long-established habits that people cease to be aware of what made them adopt the procedures in the first place (Shiffrin & Schneider, 1977). The point is that the type of exploration that is needed for different tasks can be developed with experience.

Deploying the two hands concomitantly also produced superior performance with non-verbal spatial layouts. The functions differed, in that one hand was used for the "landmark" symbol and the other for gaining information from the external surrounding frame to which the location of the landmark could be related and that could also act as a retrieval cue in recall.

The lack of a left-hand advantage in these spatial tasks, together with previous apparently contradictory findings, showed that performance by the left hand is not a reliable basis for making inferences about spatial coding. For the right-handed blindfolded participants it did not matter which hand they used for gaining and recalling information about the distances or the locations, so long as the other hand was used for gaining reference cues for spatial coding. The results are completely consistent with the neurological evidence that the right hemisphere is more involved in spatial processing, and the left hemisphere with movement or sequential information, and that both deal more with inputs from the opposite side of the body and less with ipsilateral inputs. The findings show that other cross-lateral effects, as well as spatial coding, are also involved in haptic processing of information.

Significant effects of the reference conditions had been predicted on the grounds that accuracy in tasks that require the recall of a specific distance or of a particular location depends on the reference cues that specify the distance or location in presentation and act as retrieval cues in recall. The findings are thus further support for the view that spatial coding of exploratory and scanning movements depends crucially on the reference information that is, or is made, available.

But the results also provide a reliable behavioural test of spatial coding. They showed that conditions that specify reference cues and/or alert people to their use for spatial coding improved performance, irrespective of whether the right hand or the left hand was used for the landmark symbol or the frame. The method can thus be used as a consistent test of spatial coding. It provides a reliable behavioural diagnostic tool of spatial coding, independent of hand effects that can be useful in clinical assessments.

CHAPTER FIVE

External and body-centred reference and haptic memory

The previous chapter showed that the proposed criteria can provide a test of spatial coding that is independent of hand effects in haptic location and distance tasks. However, the tests of spatial coding were based on instructions to use both egocentric and allocentric reference information, without distinguishing between them.

The results discussed in the last chapter, as well as the predictions from the theoretical description (Millar, 1994), raised the further question as to whether effects of external reference cues can be demonstrated also in purely haptic conditions, and, if so, how externally based and body-centred reference organisations relate to each other in haptic perception and memory.

This chapter therefore asks the following questions: What effect, if any, does reference information from an actual external surround have on spatial coding in tasks that depend solely on new inputs from touch and movement? If there are effects, are the processes that integrate external cues and the processes that integrate cues from body-centred sources mutually dependent, or do they operate independently of each other in haptic spatial memory?

There was no previous behavioural evidence on the effects of external reference cues in spatial coding when the external cues are also based on inputs from touch and movement. The issue is important for an adequate theoretical description of spatial coding in haptic memory, and it has practical implications for the use of tactile spatial layouts in small-scale space that can provide important information about the external environment for blind and newly blind people. It also taps into long-standing and still current differences in assumptions about how the two categories of spatial reference relate to touch.

The idea that "visual space" and "haptic space" differ, and that visual space depends on Euclidean coordinates, whereas haptic space is centred on the body and has quite different, non-Euclidean characteristics, goes back at least to the middle of the last century (e.g. Revesz, 1950). It is still sometimes assumed, at least tacitly. But the implication that haptic performance is based on body-centred reference cues and is therefore non-Euclidean, whereas visual space is based on external reference cues and consequently works according to Euclidean principles, is misleading.

An almost opposite assumption goes back at least to Katz (1925), who argued that touch puts us "in touch" with the external world. Touch actually allows us to distinguish between stimuli that arise from internal and external sources, whereas vision does not provide that information.

The reference hypothesis (Millar, 1994) links the form of spatial organisation in a particular task with the congruence of converging inputs from both external and body-centred sources that are available for processing as reference cues to specify targets in a given spatial task. Task requirements and the information that is available from different sources thus determine the type of reference organisation that is possible, not the modality as such.

Findings on both sides of the argument for and against rigid links between touch and body-centred reference, and between vision and external reference organisation, are briefly considered first.

A study that was undertaken specifically to distinguish between body-centred and external reference cues in haptic conditions is considered in some detail next. The method separates and combines body-centred and external reference cues in haptic recall of a series of spatial locations from a tactile map.

The implications are discussed in relation to haptic perception and memory for spatial information.

BODY-CENTRED AND EXTERNAL REFERENCE AND MODALITY SYSTEMS

It is not necessary to argue at length that vision is more specialised for the pick-up of external shape and spatial cues from the environment than is haptic perception. The blind conditions that tend to elicit spatial coding relative to body-centred reference cues were discussed earlier. It is not, therefore, here taken as controversial that there are links between visual conditions and externally based reference information, and between haptic conditions and body-centred reference organisation. But the links are considered to relate to the information that is available from different sources, and not as indissoluble connections between specific modalities and different forms of spatial coding.

It can also be assumed that the distinction between body-centred and external reference organisations classifies multiple inputs broadly by their source (e.g. Barrett, Bradshaw, Rose, Everatt, & Simpson, 2001; Berthoz, 1991; Paillard, 1991; Pashler, 1990; Stein 1991). In principle, therefore, any part of the body, but also any aspect of externally based cues, could provide the anchor or baseline for the reference organisation in a given task.

Errors that are due to the use of such partial, or non-optimal, baselines for spatial judgements in haptic tasks do show that the observers did not use Euclidean metrics to arrive at their judgements. The question is whether such errors necessitate the more general description of "haptic space" as non-Euclidean. A case in point is the nice study in which participants were asked to use the two hands to rotate two bars so that their orientations felt parallel each other in the horizontal plane (Kappers & Koenderink, 1999). Four different orientations were used. The standard-bars and test-bars could be positioned at nine different locations relative to each other, combining three horizontal distances from the participant's body midline in the horizontal plane with three vertical (forward direction) distances. The horizontal distances produced errors that deviated systematically from the correct settings for lateral (left–right) distances across the body midline, but not with vertical (forward–backward) distances. Further studies by Kappers and colleagues replicated the finding. The consistent deviations from parallel settings of the bars with lateral distances in the median plane were considered strong evidence that haptic distance is non-Euclidean (Kappers, 2002; Kappers & Koenderink, 1999). Kappers and colleagues (e.g. Pont, Kappers, & Koenderink, 1999) also found that haptic perception of curvature varied with the specific type of body-centred cues, either relative to the fingers or relative to the hand, that are being are used as reference anchors for the haptic inputs.

A left–right bias across the body midline has been reported by a number of studies: for instance, in line-bisection tasks. Haggard, Newman, Blundell, and Andrew (2000) tested left–right biases with hand positions in tabletop space. Positioning the tip of a pen held in the unseen hand from below the table surface at exactly the same location that was being touched by the other hand above the surface produced larger positive constant errors at left distances further from the body midline, and lower or negative constant errors at the further locations on the right of the midline, and vice versa. The bias was largest when matching unseen target locations. It was evidently not due to which hand was being used for matching.

The deviations were attributed to coding the haptic target location by the position of the hand relative to the shoulder as the basis or anchor of egocentric reference for that hand position. The authors also quote neurophysiological findings (Recanzone, Merzenich, Jenkins, Grajski, & Dinse, 1992) that show fluctuations, with relative use, in the cell populations that are activated in such tasks. Changes in cell activation that are involved in what may be called the baseline or anchor points of a particular form of body-centred reference in a given task suggest that the notion of a single, overarching body-centred body image may be less viable than is sometimes thought.

It would actually be difficult to explain all haptic errors by the notion of an unchanging body-image. For instance, changes in the length or direction of exploring movements affect haptic recall of the end, or target, location (e.g. Adams & Dijkstra, 1966; Laabs, 1973; Laabs & Simmons, 1981; Lechelt, 1982; Lederman, Klatzky, & Barber, 1985; Millar, 1979, 1981a, 1985a; Millar & Ittyerah, 1992). Such errors can be explained by heuristics based on duration, or on kinaesthetic cues when there is a lack of reference cues that define the start and end locations sufficiently. At the same time, such errors do imply that participants did not use the Euclidean principle of the shortest distance between two points.

Finding oblique effects has also been inconsistent in haptic perception (e.g. Appelle & Countryman, 1986; Appelle & Gravetter, 1985; Lechelt & Verenka, 1980). The oblique effect names the well-known finding in vision that oblique orientations produce less accurate judgements than do horizontal orientations of objects, and that vertical orientations are judged most accurately. Inconsistencies in finding oblique effects in touch might also be interpreted as evidence that haptic spatial perception differs fundamentally from visuo-spatial perception. However, the haptic oblique effect is also found in haptic conditions that provide relevant reference information (Gentaz & Hatwell, 2000, 2004).

There is indeed no question that modality-specific movement and kinaesthetic cues affect haptic spatial perception and performance, particularly in some task conditions and with some types of materials. The evidence, for instance, that the size of shapes and patterns is less easily generalised with purely haptic inputs than in vision (briefly discussed in Chapter 4) is a case in point.

However, deviations in visual perception from what are objectively parallel lines are also well known. Kappers and colleagues, among others, have shown such deviations in vision (Cuijpers, Kappers, & Koenderink, 2002). The perception of distance in two-dimensional pictures depends on the convergence of lines that are seen as parallel but are receding into the distance. Very long or very short distances, and the type of symbol being used for graphs compared to maps (e.g. Tversky, 1981; Tversky & Schiano, 1989), and even the number of landmarks along a route, can distort visuo-spatial length judgements.

There is every reason to measure performance and to score errors in terms of deviations from optimal geometrically based solutions. The covert reference organisation that can be inferred from performance errors may well show, and often does show, that people's performance in vision was not based on the use of Euclidean principles as applied to external space as defined by geometers. The inferred baseline that was used can also explain how errors occurred.

However, even though performance is scored in terms of errors relative to objective geometric axes, it is doubtful that it is appropriate or useful to identify particular types of spatial reference organisation with non-Euclidean space, or, for that matter, with Euclidean "space". Such issues differ from choosing an appropriate mathematical model to describe how either the whole system, or any part of the system, works in principle and deciding whether processes of integration require probabilistic models or description in terms of non-linear mathematics.

To identify external reference organisation with vision as "visual space" and body-centred reference organisation as "haptic space" is thus misleading. Visuo-spatial perception depends on the coordination of multisensory information, especially in relation to action. Much of that information is body-centred also in vision, in that it also involves inputs from the eye muscles, proprioceptive cues from neck and head postures, and, importantly, sensory inputs about gravitational directions from the vestibular system. The fact that body-centred cues, including proprioceptive and gravitational cues, affect vision in addition to external cues suggests that there is generally a better balance of inputs in visual conditions that can be integrated into a reference organisation that makes spatial coding more accurate. But it also means that vision cannot be identified with spatial coding based solely on external cues.

Moreover, the broad category of allocentric reference cues subsumes multiple external cues that can, in principle, provide potential reference anchors or baselines for spatial coding in different task conditions. Anchor cues or reference baselines can differ for particular locations. Similarly, the broad category of egocentric reference organisation subsumes multiple body-centred cues that can potentially be organised as reference anchors or baselines for a given spatial task. Spatial performance that relies on any one of these sub-categories alone is liable to some bias.

Findings which show that haptic shape and spatial perception does tend to be slower and less accurate than visuo-spatial perception can be explained by the greater imbalance of the information that is available in different task and/or modality conditions, respectively, from external sources and from body-centred sources. Egocentric reference organisation of inputs tends to win out in blind conditions that lack external cues.

Differences in the balance of information that is available from external and body-centred sources can also explain, for instance, why the oblique effect tends to be more variable in touch. The orientation of an object or shape depends on the reference direction to which it is related. In vision, cues for the vertical direction from body-centred gravitational cues parallel the gravitational-vertical orientations of external objects in our environment, thus providing a redundancy of information that is often lacking in purely haptic conditions.

Hearing is a "distal" sense like vision. It also provides external reference cues as well as body-centred (e.g. intensity differences with respect to the ear–head position) and proprioceptive and gravitational information. Woodin and Allport (1998) distinguished body-centred and environmentally based reference cues behaviourally by using auditory cues for external locations in blind conditions and pointing responses to indicate target locations and distractor targets near and far from the body.

Obtaining dual, seemingly "redundant", information from both external and body-centred cues – for instance, about the vertical/upright direction – is more difficult to achieve in purely haptic conditions, when these depend mainly or solely on body-centred gravitational cues. Without correction from additional information or remedial intervention, total absence of vision from birth can affect the development of the body posture. Moreover, if sighted people are blindfolded before they enter an unfamiliar environment that provides no recognisable environmental sound or smell cues, they have little other option than to rely on information from their posture and movements. The veering biases that such environments produce were considered earlier. Even a single sound from an external source may mislead in such conditions. But a sound that marks a known target or goal, and can be related to initial body-posture and "heading" cues, makes a straight-line walk possible without vision, at least to a relatively near familiar target (Millar, 1999b). Body-centred proprioceptive and gravitational cues are important in both vision and touch. But they may not always be sufficient on their own.

The link between haptic perception and body-centred spatial reference is explained here by conditions in which inputs from body-centred sources are the main or the only reliable current information that is available, when they are not balanced by information from other sources. Blind conditions are usually used in examining perception and performance by touch and movement. But there was little decisive evidence that spatial coding in haptic perception and memory may also be based on external cues derived from haptic inputs, or how such reference cues would relate to body-centred reference.

Some neurological findings that bear on the question of how the two forms of reference relate to each other are considered next.

NEUROPHYSIOLOGICAL EVIDENCE FOR BODY-CENTRED AND EXTERNALLY BASED REFERENCE

Neurophysiological studies of spatial behaviour have mostly used vision in investigating body-centred and environmentally based reference. The tasks often involve movements as outputs or actions that produce visible results.

Visual and proprioceptive feedback from reaching a target, or from positioning movements, also produce "feed-forward" effects (via corollary discharge) that affect intended actions and planning.

The findings that were discussed earlier (Chapter 2) show that functions that involve spatial processing are distributed over several regions of the cerebral cortex. These have reciprocal connections with other cortical areas and also to a number of subcortical areas. A good deal of the evidence suggests that different brain areas and circuits are involved in environmentally based and body-centred spatial tasks and that these areas are further subdivided into circuits that function in quite specific types of task requirements (Arbib, 1991; Bayliss & Moore, 1994; Berthoz, 1991; Rolls, 1991; Vallar et al., 1999).

Two main reference systems have been tested in vision in relation to each other in rhesus monkeys. Recordings from the parietal cortex of rhesus monkeys were used. In some conditions the head was rotated so that the visual information was either congruent with respect to the body axis but differed with respect to environmental reference cues, or vice versa. Responses to retinally identical targets were compared after head and body rotation, so that the head was always in line (at 0 degrees) with respect to the body but varied in angle with respect to the external stimulus location. A combination of directed eye gaze or retinal location and head position determined the location of a target relative to the body. The main findings showed that separate cell populations of neurons in the parietal cortex of monkeys are activated by conditions that tested either body-centred or world-referenced spatial coding (Snyder, Grieve, Brotchie, & Andersen, 1998). The cell populations that were activated, respectively, by body-centred and external cues occupy adjacent but separate areas in parietal regions.

The findings suggest that the two forms of spatial coding are independent. However, the authors also reported that some cells in each of the adjacent cell populations showed activation in both types of reference conditions (Snyder et al., 1998). There were very few such cells. But they could have some function in integrating body-centred and external reference cues.

Bimodal cells that respond both to visual stimuli and to somatosensory stimuli in the macaque putamen, in otherwise modality-specific neuron populations, evidently also contribute to multisensory integration in spatial tasks (Graziano & Gross, 1993). These bimodal neurons that are activated by somatosensory and visual inputs are considered to provide a "map" of space immediately surrounding the body, as part of a defensive system that alerts the animal to closely approaching objects and/or prepares for immediate defensive action. It suggests a relatively closed, special-purpose system. That may differ from what can be considered personal space, or space within reach, or indeed for space beyond reach, by using canes or other tools, when the action is not potentially defensive, but on the contrary one of actively gaining information from the environment for other purposes.

Post-parietal regions in dorsal stream processing of spatial information in the human cortex are homologous with parietal regions in the monkey brain. In humans, non-invasive fMRI studies have shown that body-centred spatial coding in both vision and touch activates a bilateral fronto-parietal network, though more extensively in the right hemisphere (Galati, Committeri, Sanes, & Pizzamiglio, 2001; Galati et al., 2000). Indeed, spatial locations can be represented in the human brain with respect to different classes of reference bases that are either relative to, or independent of, the participant's position and may be either object-based or centred on a body-part. But judging the location of the visual stimulus by reference to external, object-based cues affected a less extensive unilateral region (Galati et al., 2000). Hippocampal regions in human and other primate brains, as well as in the rat brain, are involved in short-term memory for allocentric visual locations for navigation (Bayliss & Moore, 1994; Rolls, 1991).

Specialisation in different regions and sub-regions for different tasks is well documented. A case study showed that bilateral hippocampal damage impaired memory for allocentric inputs more than for egocentric information (Holdstock et al., 2000). There is also increasingly persuasive evidence for the importance of connecting circuits. Hippocampal regions receive inputs from post-parietal cortical areas and are severely affected by insults to the vestibular system. Integration of diverse stimuli for action evidently involves various levels of the central nervous system (Berthoz, 1991, 1993; Maioli & Ohgaki, 1993; Paillard, 1991). Paillard (1991) argued that gravitational information from the vestibular system is "geo-centric" and, as such, is a common factor in allocentric and egocentric forms of reference.

The neurological picture underlying egocentric and allocentric spatial processing in vision thus seems typically one of specialisation for greater diversity of information, on the one hand, and multiple integrative processing, on the other. The biological advantages of specialisation would presumably be lost without integrative organisation. Whether that also necessitates an overarching general spatial representation is still in question.

The two forms of reference had been isolated in visual and in auditory tasks. However, despite the general assumption that spatial coding in purely haptic conditions depends solely on various forms of body-centred reference, there were some indications from behavioural observations that cues external to the stimulus information can also be used in purely haptic conditions. The observations are considered below.

INDICATIONS FROM OBSERVATION

Incidental observations of fluent braille readers suggested that, with experience, they do use externally based spatial coding. Early in learning, aligned head and body postures and body-midline cues are extremely important for learning to read braille. However, posture cues are much less important for fluent braille readers. The experiments with rotated texts described earlier (Chapter 4) showed that it was far more important for fluent readers to keep the lateral shear patterns on the finger orientated to the rotated external text, despite the awkward change in body-posture that such lateral scanning required, than to keep the finger position in line with an upright body-posture (Millar, 1987a, 1997).

Interestingly enough, some of the best – and most cooperative – braille readers systematically ignored instructions to wait until the start signal before touching the text in reading for comprehension. Unaware that their hand movements could be viewed on the monitor prior to the starting signal in experiments, they consistently used both hands to explore the rectangular outlines that the raised layouts of braille texts produce.

The fact that such initial exploration of the physical shape of the layout did indeed provide useful spatial information was evident from other braille findings. Fluent braillists can find a specific word at earlier locations in a text, in order to disambiguate its meaning or to comprehend an indistinct term to which it seems related. In one study, simple homophone words were embedded in appropriate and inappropriate contexts at different locations in the same passage. Other studies tested the use of semantic context to comprehend faint, physically degraded, braille words by embedding them in prior or subsequent semantically appropriate sentences in the text. It was clear from the regression movements to different parts of the text that fluent braille readers were able to locate the relevant critical word or context quickly and accurately within the spatial layout of the text (Millar, 1988c, 1997).

The point is that the usual rectangular layout of braille texts provided an actual external spatial frame that made it possible to locate critical words within the layout of the text easily by touch. Such strategies evidently required a good deal of experience. But they suggested that the external surrounds of relevant stimuli may provide external reference cues also in purely haptic conditions.

According to the reference hypothesis, there is actually no reason to suppose that haptic conditions prevent people from relating cues from felt external surrounds to target locations, or prevent them from using that information as retrieval cues that allow the target to be located later. There is a tendency to code object locations by reference to the body-midline and body-posture cues in haptic conditions for reasons discussed earlier. But it should be possible to code locations in tabletop space by reference to external cues that surround stimulus objects also in touch.

Specifying the location of objects relative to each other, or relative to a surrounding frame, does not necessarily have to depend on egocentric reference alone. One's pen can be located with respect to a corner of the table, or relative to the known position of a nearby lamp, rather than to body-centred cues, even if the hand is employed in the exploration.

However, there was no direct experimental evidence on whether effects of external cues on spatial coding can be demonstrated with purely haptic inputs and, if so, how the two types of coding spatial information relate to each other in haptic perception and memory.

The questions are of some importance for methods of rehabilitating newly blind adults, as well as for theoretical accounts of how spatial processing of haptic inputs takes place. The experimental manipulations that were used to investigate whether providing external reference cues would also affect haptic spatial coding, and, if so, how egocentric and allocentric coding relate to each other, are therefore described in some detail below.

TESTING EXTERNAL AND BODY-CENTRED REFERENCE IN HAPTIC TASKS

The experiments (Millar & Al-Attar, 2004) discussed here tested predictions from two hypotheses. The reference hypothesis predicts that the form of spatial reference depends on what information is most readily available and reliable in given task conditions (Millar, 1994). The hypothesis that spatial coding divides along strictly modality-specific lines implies that haptic processing of spatial information depends solely on coding targets in terms of body-centred cues.

Experimental evidence for the predictions was sought by using set-ups that pitted the two types of reference information against each other in tasks that were based on purely haptic inputs and performance. Vision normally provides external background cues for spatial targets. Moreover, their use for reference in visuo-spatial coding and memory is often automatic, and it rarely needs explicit attention. Visuo-spatial cues were therefore excluded. The volunteers who participated were blindfolded before entering the experimental space, and they remained blindfolded throughout.

Task conditions that exclude external cues from vision do make it necessary to alert people explicitly to external cues from unfamiliar haptic external surrounds of targets and to their use as reference information for target locations. That does, of course, constitute a difference in modality conditions between vision and touch. But the hypothesis that haptic targets can only be coded spatially in relation to body-centred cues would predict that providing haptic cues explicitly from an external surround does not improve recall accuracy beyond the level found with body-centred reference cues alone. If, on the other hand, the difference in spatial coding is due solely to the lack of external reference information that is normally available in new haptic task conditions, providing external haptic cues explicitly for reference in a spatial task should improve recall significantly.

We deliberately chose a spatial task that people might actually encounter. Participants were to remember the precise locations of "landmarks" en route to a goal or end location from an embossed tactile map-like display. Embossed tactile maps are, in fact, used by blind adults, although tactile maps are rarely produced with surrounding frames. The omission seems to stem from the belief that external-frame information is less, or not at all, useful in blind conditions. The map-like layout we used had an actual, tangible, rectangular surrounding frame (Figure 5.1). But the frame was only brought into play in the conditions that tested specifically for effects of external cues on recall accuracy.

The task was to remember the precise location of five shape symbols as "landmarks" that had been positioned randomly along an irregular, but easily felt, raised-line route (Figure 5.1). The "landmarks" that marked the five target locations were distinctive small raised shapes. The shapes were easy to discriminate and were further practised in pre-test conditions that preceded the trials. In presentation, the layout was always placed on the table and aligned to the participant's body midline.

Participants were helped to place the fingertip of the preferred (right) hand at the start of the route in all conditions. The experimenter briefly

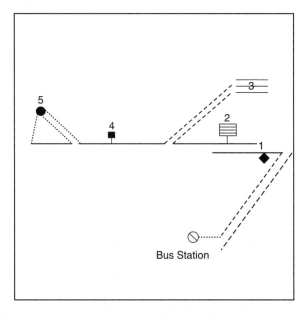

Figure 5.1 The designated route in the presentation map with symbols as landmarks for the to-be-remembered locations along the route. [From S. Millar & Z. Al-Attar, "External and body-centred frames of reference in spatial memory: Evidence from touch", *Perception & Psychophysics*, 2004, *66*, 51–59. Copyright © Psychonomic Society Inc.]

guided the scanning finger in the correct direction. Participants then scanned the route from left to right in all presentation conditions. They were told to stop briefly on each landmark symbol that they encountered on the route. They were to place the midpoint of the finger pad on the midpoint of each of the five landmark symbols. Participants were told that these were the precise locations that were to be remembered for the recall tests.

The same map layouts were used in recall, except that they did not contain landmark symbols for the to-be-remembered locations. In all recall tests, participants were to place the midpoint of the pad of the scanning finger on the remembered midpoint of the landmark symbol in turn for each of the five targets. The experimenter marked the midpoint of the fingertip in recall for each location on the grid lines of ruled replicas of the map layouts. The marks showed the positioning (mm/cm) errors with respect to the correct location for each of four trials of the five target locations in every reference condition.

There were four reference conditions. In the body-aligned control condition, the map was aligned to the participant's body midline in tabletop space for scanning the landmarks. The test map did not contain the landmark symbols. But the test map was aligned to the participant's body midline in exactly the same way. The five to-be-remembered locations would thus be in exactly the same position relative to the body in recall as in presentation. Body-centred cues were thus available as reference cues for the locations in presentation and for retrieval in recall. To test body-centred coding, the presentation map with landmark symbols for the locations was aligned to the participant's midline as before. But the test map without the landmark symbols was rotated in the recall test, so that body-centred coding of the locations was disrupted. The frame surrounding the map was not touched or mentioned at any time in either the body-aligned control condition or in the rotation conditions. There were thus no external cues in these two conditions.

Map rotation was used to disrupt body-centred coding. The presentation phase was precisely the same as in the body-aligned conditions. But the recall map was rotated clockwise by 90 degrees relative to the body midline. Rotating the display with respect to the person in recall is one of the most common means of disrupting body-centred reference. But there is a difficulty in using rotations. Rotating a display during brief delays tends to increase memory load by eliciting mental rotation or other cognitive strategies to update target locations. That difficulty had to be avoided. The only point of the rotation here was to disrupt egocentric coding, in order to compare it with accuracy when body-centred coding remained intact in recall. The new direction of the route in recall was, therefore, made obvious. The scanning finger was guided in the correct direction for an initial, small part of the route in all conditions, as noted earlier. That avoided the need to remember the direction of the route in any condition. More important, it also meant that the new direction of the route after rotation did not have to be computed. Only the locations of the five landmarks on the route had to be remembered in any condition. The difference in positioning errors under rotation compared to intact body-centred coding thus measured the use of egocentric reference cues.

To isolate the use of external reference information, body-centred reference was disrupted by map rotation as before. But participants were alerted to the presence of the actual tangible rectangular border that framed the map layout and were encouraged to feel around it in the pre-test conditions. Participants were explicitly instructed to feel the frame with the free hand both in presentation and in recall, and to use information from the frame to specify and to remember the location of the felt targets relative to the frame.

The fourth reference condition combined intact body-centred reference with external reference-frame information. The map-to-body orientation remained the same in recall as in presentation. In addition, the participants were instructed, as before, to use information from the surrounding frame relative to the target locations in presentation and in recall.

Thus, in the aligned condition there were no external cues, but body-centred cues were the same in recall as in presentation. The second condition also lacked external cues and also disrupted body-centred coding. The third condition provided external reference information from an actual external frame that surrounded the to-be-remembered location along the designated route. But body-centred coding was disrupted by rotating the map, so that only external reference was reliable in recall. The fourth condition provided both allocentric and egocentric reference information in presentation and recall. Forty intelligent young adult volunteers were allocated randomly to the four reference conditions and tested blindfolded.

Predictions from the two hypotheses under test differ for three of the relevant comparisons. Both hypotheses assume that disrupting body-centred coding would increase positioning errors for the locations compared to the aligned condition. The hypothesis that spatial coding of haptic inputs is linked solely to body-centred reference frames implies that adding external reference information from touch and movement should have little or no effect. Adding external reference cues should, therefore, not reduce the positioning errors found in conditions that disrupt body-centred coding and should not increase the level of accuracy beyond that provided by intact body-centred reference. The hypothesis that spatial accuracy depends on the convergence of inputs that are potentially available from any source predicts that added external reference information reduces errors significantly, even when body-centred coding is disrupted. Adding external reference cues should also improve spatial accuracy significantly further when it is combined with intact body-centred reference.

The results clearly supported the second hypothesis. Statistical comparisons of the four reference conditions showed highly significant effects.

Disrupting body-centred cues by rotation increased errors significantly compared to intact body-centred coding in the body-aligned condition, as expected on both hypotheses. The critical results were, first, a highly significant decrease in positioning errors with added external reference information when body-centred coding was disrupted by rotation, compared to the rotation condition that lacked external reference information. The condition with intact body-centred cues and added external reference information was very significantly more accurate in comparison to the body-aligned condition without external cues, and very significantly more accurate also than the condition with added external information when body-centred coding was disturbed by rotation. Furthermore, accuracy with added external information, but disrupted body-centred coding, did not differ from intact body-centred coding without external reference information (Figure 5.2).

The experimental manipulation of separating and combining external and body-centred reference thus showed that external reference cues can also be used with purely haptic information. Indeed, when external information is made available in haptic conditions, it seems to be equally as effective for spatial coding as is body-centred reference information.

Perhaps more important for understanding the relation between external

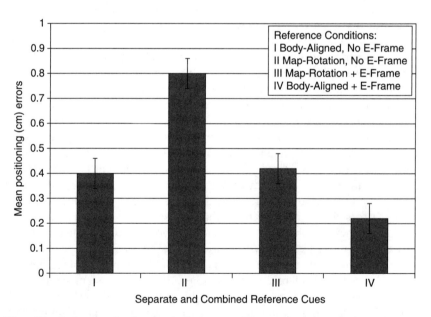

Figure 5.2 Body-centred and external reference cues. Mean errors and standard errors suggesting additive relations between reference conditions. [From S. Millar & Z. Al-Attar, "External and body-centred frames of reference in spatial memory: Evidence from touch", *Perception & Psychophysics*, 2004, *66*, 51–59. Copyright © Psychonomic Society Inc.]

and body-centred reference in haptic perception and performance was the relative size of recall errors. When body-centred coding was disrupted, the presence of external reference information produced the same level of accuracy as body-centred coding in the absence of external cues. External reference without intact body-centred reference was thus as effective as body-centred reference without external reference. Furthermore, when both intact body-centred cues and external reference cues were available, errors were significantly halved compared to either of the single reference conditions. The findings thus suggest additive effects between the two reference conditions. Theoretically, additive effects imply that the two forms of spatial organisation are independent of each other.

It should be noted that the relation between errors in different conditions also speaks to the question of whether effects for blindfolded sighted participants could have depended on the use of visual imagery. They had all been told why we were using a map-like tactile layout. On the assumption that participants translated all haptic information into visual form, there would be no reason to expect cumulative or independent effects of the two reference conditions. Alternatively, translation into visual imagery may be more difficult with body-centred reference information than with external reference information. If so, the external reference condition should be more accurate than the body-aligned condition. The findings showed neither of these outcomes. They were consistent instead with the example described earlier of the use made by blind readers of the external layout of a braille text in locating critical words within it, which partly prompted the experiment.

The results were quite clear with regard to the two hypotheses. It would not be possible to demonstrate the independence of external and body-centred coding in purely haptic conditions if spatial coding were divided along strictly modality-specific lines. The findings showed that relating target locations to a surrounding frame when that information was made available was as effective as relating the target locations to body-centred cues.

An interesting, and unexpected, finding needs to be mentioned. The placements of the five locations along the route had been intended to be random. But they produced a significant effect in all reference conditions. The effect was principally due to greatly reduced errors for the final location and, to a lesser extent, to reduced errors also for the first location in contrast to the middle locations. Serial-order effects that produce such quasi-bow-shaped curves are typical in the recall of verbal serial items. But serial-order effects have also been found in visuo-spatial tasks (e.g. Jones, Farrand, Stuart, & Morris, 1995; Smyth, Hay, Hitch, & Horton, 2005; Smyth & Scholey, 1996). The quasi-bow-shaped curves found here could not be attributed to a verbal serial position effect, nor even to the need of recalling the order of items. In order to be able to rule out such effects, the method here required the experimenter to name each landmark before it was encountered in presentation and also before the participant indicated the location in recall. However, the final location had inadvertently been placed at the end of the route. The change in touch from the raised texture of the landmark and end of the route to the smooth surface beyond it thus provided an additional local cue for the final target location. The first target was near a distinctly sharp bend in the route (Figure 5.1). The layout of the route thus suggested that the location effect was due to the fortuitous touch cues near the first location and at the final location.

The comparison of the body-aligned and rotation conditions also produced a small but significant interaction in addition, due to a larger difference between the rotation and aligned condition at the first and the middle location than at the other locations. A possible role of disrupting the midline alignment in map rotation specifically for these locations is discussed further in conjunction with the different location effects found with a quite different spatial layout (see Chapter 8). In both cases, the location effects suggested that salient touch cues near target locations acted as additional retrieval cues for recall. Such local touch cues seem analogous to object- or shape-based reference cues that are considered further in the following chapters.

PRACTICAL AND THEORETICAL IMPLICATIONS OF THE FINDINGS DISCUSSED IN THIS CHAPTER

As far as I know, allocentric coding had not previously been demonstrated and distinguished experimentally from egocentric coding in behavioural studies with purely haptic inputs. Separating and combining the two forms of reference showed that when information from an actual external frame outside target locations was made available, it improved haptic recall significantly, as predicted by the reference hypothesis. Indeed, external-frame information alone improved haptic location recall as much as intact body-centred information alone, and the combination of the two forms of reference information produced twice the level of recall accuracy compared with either form of information on its own.

The findings are not compatible with the view that haptic inputs can only be coded spatially by reference to body-centred cues and that the use of external reference cues requires inputs from vision or some other distal modality. Visual conditions do differ from tactual conditions in the extent to which they provide external background cues routinely. However, when cues from the external reference frame beyond target locations were known to be available in haptic conditions, they added very significantly to recall accuracy.

There are practical applications that need to be taken into account by health professionals who are concerned with rehabilitation. Tactile maps that provide information about a route and landmarks in an external environment, and graphic tactile layouts of tools and necessary gadgets, can be extremely useful in conditions of visual impairment. But tactile maps, unlike visual maps, are rarely furnished with frame surrounds that can provide coordinate cues that specify locations.. The results here suggest, first, that information from an external frame can also be used in the absence of vision and, second, that external information can be as effective as body-centred spatial reference. For inexperienced people, the use of surrounding frames for reference information may need to be explained and signalled explicitly. Above all, the fact that recall accuracy doubled when information from the external frame was combined with body-centred reference needs to be stressed. The apparent "redundancy" of reference cues for target locations is important when the procedures are not yet habitual. Such combinations of body-centred and external cues, and explicit instructions on how to use these for either visual or haptic targets in contralateral space, may also prove useful in rehabilitation in conditions of unilateral neglect.

Theoretically, the findings have several implications. They are consistent with the predictions from the reference hypothesis and with the general description of spatial processing as the organisation of inputs from diverse sources in terms of reference information for location and distance tasks. Moreover, the additive nature of the integration of reference information from body-centred and external sources, which suggests that they had independent effects, is also consistent with the neurophysiologial evidence from studies with vision, briefly outlined in Chapter 2. The areas of the brain that are involved in spatial processing are widely distributed, and, even within the inferior parietal cortex, adjacent but mainly separate populations of neurons seem to be involved in egocentric and allocentric task conditions. Behavioural evidence that egocentric and allocentric coding are independent for haptic inputs, as they appear to be for vision, is interesting. It seems to suggest that the "rules" of integrating inputs for different forms of reference are the same for the two modalities.

At the same time, "top-down" effects are likely to be more involved in purely haptic recall conditions, if only because people have to be alerted to the presence of external cues, unless the contexts are familiar and they have learned by experience to expect external cues, and know how to search for them. "Feed-forward" from experience is presumably involved in planning the pick-up of external-frame cues with one hand and using the other hand concomitantly to detect the target so that its location can be coded relative to frame cues. That was clearly not necessary in the short-term memory conditions described here, in which the observers were explicitly instructed that an external frame was present and how it should be used.

A further implication concerns distinctions that need to be made within what may broadly be described as short-term memory for touch and movement. Haptic inputs from a felt landmark were coded relative either to body-centred cues, or to adjacent or surrounding cues, or to both. The presence of cues that can specify the location of a target is even more important for finding the location again in the absence of the landmark after a short delay.

Short-term memory for spatially coded inputs from touch and movement inputs that was demonstrated here needs to be distinguished both from recall of serial tactual dot patterns and from memory based mainly or solely on kinaesthetic inputs. The very small recall spans for serial dot patterns that depended on touch alone as contrasted with spans for nameable patterns (Millar, 1975b) were mentioned earlier. A further distinction is needed for memory for kinaesthetic information from movements, as such. The difference, shown by a right-hand advantage for longer and more complex recall movements that did not interact with reference cues, was described in the study on hand movements in the previous chapter. Spans for a sequence of different movements that are felt without vision are larger than for serial tactual dot patterns (Ballesteros et al., 2005). It is also possible to rehearse simple movements covertly during short delays in the total absence of vision (Millar & Ittyerah, 1992). But spatial coding of haptic information, as shown by the separate and combined effects of different forms of reference cues in the study described in this chapter, adds a further factor. It suggests that short-term haptic recall of spatial information should be distinguished from purely tactual recall and from short-term memory for kinaesthetic inputs.

The characteristics of haptic short-term memory for spatial information are analogous to distinctions made in visuo-spatial tasks, although they may not parallel these exactly. Studies of visuo-spatial coding have led to a considerable enlargement of the inputs and functions described in terms of the "visuo-spatial sketchpad" in Baddeley's later (2007) model of working memory. It includes kinaesthetic inputs. Sequences of movements that have been viewed may be rehearsed covertly and so may play a rehearsal role in maintaining visual information across short delays with little loss of information. Spatially relevant pointing, arm movements and spatial tapping have been shown to interfere with visuo-spatial memory (Della Salla, Gray, Baddeley, & Wilson, 1999; Lawrence, Myerson, Oonk, & Abrams, 2001), suggesting a complex storage system that also dissociates memory for locations from visual memory for objects. An additional episodic buffer (Baddeley, 2000) was assumed as a temporary storage system that integrates information from different dimensions into coherent episodes. A central control system, reciprocally connected with the subsidiary systems, is concerned in the capacity limits in different systems. The different subsystems also connect reciprocally (Baddeley, 2007).

Another distinction in vision that was also found in touch is the significant effect of local touch cues as a means of reference. It was shown by the location effect, based on errors for specific locations. It showed that people use salient local touch that happened to occur near a target location as anchor or reference points in addition to body-aligned and/or external reference cues.

The use of such apparently irrelevant local touch cues may contribute to task-specific effects. However, similar effects of local cues and of actual square frames are also found in visuo-spatial tasks (e.g. Igel & Harvey, 1991).

Similarities and differences between haptic and visual perception and memory for spatial information constitute some of the potentially most telling pointers to the relation of modality-specific inputs to spatial integration.

The next two chapters look at similarities and differences between haptic and visual shape illusions.

CHAPTER SIX

Visual illusions that occur in touch

Visual "illusions" are intriguing examples showing that what we see is not always what the ruler measures. The great nineteenth-century physiologist Helmholtz (1867, 1896) held that illusions are due to discrepancies in the very cues that normally produce accurate perception. Once we can explain what information produces accurate perception, misperceptions simply become an aspect of how we see. It seems to me that the principle proposed by Helmholtz is still valid.

Optical illusions have produced large numbers of investigations. Some illusions, such as discrepancies in motion, fusion and time intervals between light stimuli, had to be investigated in detail before motion pictures and television viewing could become tolerable, let alone enjoyable. There is no general consensus as yet on how illusions should be explained. But the study of ambiguous figures, of backgrounds that change the colour or appearance of an after-image and of straight lines that look curved has added to what we know about visual perception.

Paradoxically, "optical" illusions have also long been known to occur in touch by blind people (Bean, 1938; Burtt, 1917; Révész, 1934). The present chapter is solely about two such "optical" illusions that occur in touch. They are produced by quite simple shapes. But they have attracted apparently definitive, but quite opposite, conclusions. Not the least controversial is the view that the similarity of tactual illusions to visual illusions is simply fortuitous.

Optical illusions that are also found in touch raise the wider question of how shape perception by vision and by touch relate to each other, and what, if anything, there is in common between the illusions in the two modalities in either case. Clearly, the receptor systems in vision and touch differ radically. At the same time, as we saw earlier, there is incontrovertible evidence that shape inputs from vision and touch have crossmodal effects. The fact that crossmodal effects occur does not, of course, itself explain similarities in illusions. But it does raise the question of whether it is appropriate to assume that it is simply a matter of complete chance that the same shape which produces an illusion in vision produces the same illusion also in touch.

This chapter discusses some of the evidence that has produced diverse explanations for the vertical–horizontal illusion. By contrast there has been little or no explanation for the quite different illusion with which the vertical–horizontal illusion is often confused, namely the bisection illusion.

The vertical–horizontal illusion names the tendency to overestimate the length of a vertical line relative to a horizontal line in L-type configurations in which the constituent lines are objectively equal. The bisection illusion occurs in T-type shapes in which the bisecting line is overestimated relative to the bisected line. Both illusions have been found also in touch.

It is suggested that, far from all explanations of the tactual vertical–horizontal being implausible, most pinpoint a factor that may indeed contribute to the illusion to some extent. That includes modality-specific effects, including differences in movement information (Gentaz & Hatwell, 2004; Heller, Brackett, Salik, & Scroggs, 2003). But modality-specific differences do not necessarily exclude the possibility that the main basis of an illusion that occurs both in touch and vision is a factor that is in common between the visual illusion and the tactual illusion.

Experiments are described that had been designed specifically to test some of the explanations that have been given for the vertical–horizontal illusion and to establish whether the similarity to the visual illusion is explained by a factor that obtains both in vision and touch. Experiments with the bisection illusion are discussed separately.

Experimental details are given for experiments that tested the Helmholtz (1867, 1896) principle which is assumed here, namely that perceptual illusions are due to discrepancies in the very cues that normally produce accurate perception. The reference hypothesis assumes additionally that the accuracy of shape perception in both touch and vision is due to convergent reference cues. The implication is that perceptual illusions that occur both in touch and in vision are due to discrepancies in the spatial reference information that normally produces accurate shape and spatial perception in both vision and touch for a given configuration.

It is assumed further that discrepancies in the type of reference cues may vary for different illusions. Discrepancies may occur relative to external cues or to body-centred cues. Discrepancies could also be due to object- or shape-based reference cues. In that case, discrepant cues from constituent features within a configuration would produce the illusion.

The findings suggested that the haptic vertical–horizontal illusion can be attributed to discrepancies in the spatial context to which the vertical and horizontal lines in a shape are related, consistent with at least one important explanation of the visual illusion.

The bisection illusion had been demonstrated previously in vision and in touch. But it had not previously been explained. The findings here suggest that the overestimation of the bisecting line occurs because it is both seen and felt as a continuous line that is bounded only by the start and end location of the line. The bisected line, by contrast, contains the bisecting point in addition. That point acts as a boundary that divides the bisected line into two small lengths. The effect of scanning the smaller sections produces the overestimation of the continuous line. A test of that hypothesis showed that the bias increases inversely with the length of the final portion of the bisected line.

It is argued that the Helmholtz principle holds despite modality differences, but that the form of reference information that is out of kilter differs for different illusions. The reference point that produces the imbalance in length cues in the bisection illusion differs from the reference cues that produce the discrepancy in length information from movements that produce the vertical–horizontal illusion in touch. But the discrepancy in each case also explains the same illusion in vision.

THE VERTICAL–HORIZONTAL ILLUSION

The vertical–horizontal illusion was reported as an "optical" illusion in the mid-nineteenth century (Fick, 1851). The original observation was that a bright square on a dark ground looks like a vertical oblong. The phenomenon was examined by both Helmholtz and Wundt.

Old and famous though it is, the visual vertical–horizontal illusion should come with a health warning. The overestimation, when it occurs, is of the length of vertically oriented lines or objects in 3-D or in 2-D (tabletop) space relative to horizontal lines or objects, typically exemplified by L-shapes. The illusion is not found when the vertical and horizontal lines do not meet. The vertical–horizontal illusion has often been tested mistakenly with upright or inverted T-shapes, which consist of bisecting and bisected lines. Overestimation of the vertical line is actually much more invariant and stronger when tested with T-shapes or inverted T-shapes than it is for L-shapes.

In vision, the bisection illusion has been distinguished from the vertical–horizontal illusion (Finger & Spelt, 1947; Howard & Templeton, 1966; Künnapas, 1955a, 1957a, 1957b). Figure 6.1, drawn after an illustration by Künnapas (1955a, 1955b), shows rather convincingly that vertical lines are by

122 SPACE AND SENSE

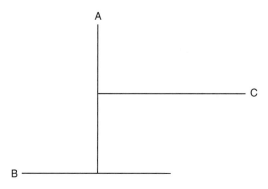

Figure 6.1 Illustration of bisecting and bisected vertical and horizontal lines of equal length.

no means always seen as longer than horizontals. Künnapas was one of the surprisingly few researchers who kept the bisection and vertical–horizontal illusions apart.

Explanations of the vertical–horizontal illusion originally centred on vision (von Collani, 1985). Some hypotheses were clearly influenced by shapes that include bisecting lines. One hypothesis, for example, explained the illusion in terms of size–distance constancy or perspective rules that govern the perception of 3-D objects and are mistakenly applied to 2-D shapes. Size–distance and perspective effects have an important place in explaining the perception of shapes that are actually tilted, or which appear to be tilted in 2-D pictures with frame and pictorial depth that are used to simulate perspective effects. It is possible that this contributes to the illusion (Gregory, 1990). Pictorial depth and framing effects evidently have separable effects (Williams & Enns, 1996). However, constancy or perspective rules are unlikely to be the main factor in the vertical–horizontal illusion as such (von Collani, 1985; Heller et al., 2003; Wolfe, Maloney, & Tam, 2005).

Künnapas (1955a, 1955b, 1957a, 1957b, 1957c, 1958, 1959) should probably be credited with the most credible explanation of the visual vertical–horizontal illusion and the best empirical evidence for it (Prinzmetal & Gettleman, 1993). Künnapas argued that the relative overestimation of vertical compared to horizontal lengths is due to the roughly elliptical shape of the visual field. A vertical line occupies a much larger portion of the vertical extent of the visual field than an equivalent horizontal line, which occupies a much smaller portion of the longer horizontal extent of the elliptical visual field (Künnapas, 1957a). Length estimation of a line is thus proportionate to the spatial frame in which the line is seen.

There is evidence for several of the predictions that follow from the hypothesis proposed by Künnapas. Changing the boundaries of the visual field by presenting a lighted configuration in the dark was found to reduce the overestimation (Künnapas, 1957a). But that reduction has not always been found (Avery & Day, 1969). However, it is possible that the smaller extent of vertical compared to horizontal eye movements has similar proportional effects on vertical and horizontal length judgements in the dark as does the elliptical eye field in vision. There is also evidence that the distance of a line from its retinal boundary determines its subjective length. Vertical lines are nearer to the vertical extent of the visual field, and horizontals are further from the elliptical boundary. Length estimations varied with the visual field as bounded by the nasal line. The prediction that the extent of the illusion differs with the configuration of the field was borne out when tested with the elliptical shape of the binocular field compared to the less asymmetric monocular fields that are comparatively truncated horizontally (Künnapas, 1957b).

The reduction of the vertical overestimation with monocular viewing was replicated (Prinzmetal & Gettleman, 1993). Spectacles that artificially extend the field of view in the horizontal direction also affected length judgements, as predicted, although other influences could not be ruled out (Künnapas, 1959).

Among the additional effects must be reckoned the set of length inputs that are used to test the illusions in the context of the task. There is good evidence that the range of lengths that are used as comparison stimuli in psychophysical judgement tasks have clear effects on judgements (Armstrong & Marks, 1997). Some findings suggest that such context effects are specific to the eye and the region of the retina being stimulated (Arieh & Marks, 2002). But such context effects on judged length, from the set of lengths that are used for comparison, also occur in touch and have much the same effect (see bisection effects later).

Primarily, however, the visual vertical–horizontal illusion was explained as a purely optical illusion, due to the elliptical retinal visual field. Inclining the head by 90 degrees changed the overestimation of verticals into an underestimation (Künnapas, 1958). Avery and Day (1969) and Day and Avery (1970) produced cognate findings. Day and Avery (1970) argued explicitly that the vertical–horizontal illusion is specifically visual. They predicted that it would not occur in touch and, indeed, failed to find any vertical overestimation for the haptic L-shape in their own study. But their L-shape was presented in the fronto-parallel (upright) plane.

Hatwell (1960) presented haptic L-shapes in (flat) tabletop space to blind children and also failed to find the illusion in touch. But the blind children showed the bisection illusion. It should, however, also be said that the L-shape illusion is not always found in vision either. Reid (1954), for instance, failed to find the visual illusion when he asked people to reproduce vertical and horizontal lines that they viewed at a distance of 6 ft on a fronto-parallel-oriented screen by drawing the lines with the unseen (screened) hand.

Reid (1954) demonstrated a haptic illusion for L-shapes in tabletop space that most people consider at least complementary to the vertical–horizontal illusion in vision. Blindfolded sighted people traced with a stylus through an L-shaped slot carved in a square placed on the table in front of the participant. The initial standard was either vertical or horizontal. It had to be reproduced after a 90-degree turn in the plate, either to the right or to the left after a vertical L-standard was presented, or up or down after a horizontal standard was presented. Vertical movements, towards and away from the body, were significantly overestimated with respect to horizontal movements to the right or left across the body. Reid speculated that vertical lines take longer to draw.

The movement-speed explanation of the haptic vertical–horizontal illusion was subsequently tested by a number of researchers. Overestimation of vertical lines was attributed to speed differences between radial and tangential movements. Vertical movements in the mid-transverse plane towards and away from the body involve radial movements from the elbow. They were assumed to be slow because they produce greater resistance to inertia (Wong, 1977). Horizontal extents require tangential movements, initiated from the shoulder. The movements swivel around the body and are faster (Davidon & Cheng, 1964; Day & Wong, 1971; Deregowski & Ellis, 1972; Wong, 1977).

Interestingly enough, some of the experiments that employed methods that produced different results varied the angle of the board on which L-shapes were scanned, relative to the flat tabletop plane, or tested outstretched arm movements at shoulder height. The movement hypothesis was considered to explain why no haptic L-illusion was found in the vertical (fronto-parallel) plane (Cheng, 1968). It was suggested that vertical as well as horizontal movements of the outstretched arm at shoulder height require only tangential movements (Wong, 1977). However, these findings also imply that the illusion differed with the spatial plane in which it was presented.

Marchetti and Lederman (1983) found no evidence to support the (Wong, 1977) movement-inertia hypothesis. Altering movement inertia by changing the distance of the hand from the axis of rotation had no effect, suggesting that the illusion cannot be explained by simple biomechanical factors. Moreover, the movement-speed hypothesis alone cannot explain why the haptic L-illusion has not always been found in tabletop space either (Hatwell, 1960; Heller & Joyner, 1993), in contrast to the inverted T-shape illusion that Hatwell (1960) demonstrated in young blind children.

Heller, Calcaterra, Burson, and Green (1997) found a haptic L-illusion when arm movements from the shoulder were allowed, but no L-illusion for very small stimuli or in conditions that immobilised the arm so that arm movements from the shoulder were prevented. Exploring when the elbow was lifted also affected the illusion. Heller argued that the haptic vertical–horizontal illusion is explained by radial versus tangential movements – that is to say, by the difference in scanning movements that involve the shoulder or the lifted elbow compared to finger movements when the arm is immobilised – and concluded that the similarity of the haptic to the visual vertical–horizontal illusion is a chance phenomenon.

The length and direction of scanning movements may well play a role in the haptic vertical–horizontal illusion. But differences in the extent of movements that involve the shoulder compared to movements that involve the elbow joint, or cues from the hand or other fingers, also have to be considered in terms of the spatial relation of the movement extents to the shoulder, or to other body parts as reference points that delimit the movement, as for instance, in the left–right, right–left bias (Haggard et al., 2000) that was discussed in the last chapter. Similarly, very small stimuli rarely produce the L-illusion in touch. But the short scanning movements that are possible with the thumb, or with the forefinger when the arm and wrist are immobilised, would make it difficult to distinguish the extent of movements relative to different body-centred cues.

Movement and possible spatial effects thus needed to be checked out in testing the radial/tangential hypothesis.

TESTING THE RADIAL/TANGENTIAL MOVEMENT HYPOTHESIS

The reference hypothesis implies that haptic illusions may well involve discrepancies in information from movements – that is to say, discrepancies in cues that are modality-specific. But it also predicts that discrepancies in the reference information by which movement extents are coded spatially are major factors in the illusion. The explanation proposed by Künnapas for the visual vertical–horizontal illusion can be applied to the haptic vertical–horizontal illusion. Discrepancies in the spatial field, or the reference cues it affords, determine the illusion not only in vision but also in touch. The spatial reference, or spatial field to which the movement inputs relate, rather than the movement inputs alone, would thus explain why and when the illusion occurs in touch as well as in vision.

However, the movement hypothesis needed checking out. It was not always very clear from previous studies how one can be certain that scanning a vertical line involves only purely radial movements from the elbow, or that scanning a horizontal line depends solely on tangential movements from the shoulder.

It seemed reasonable to assume that the relation between radial and tangential movements differs when scanning a shape by the right or left hand in ipsilateral (same side) hemispace as compared with scanning the same shape in contralateral (opposite side) hemispace. If, as the radial/tangential hypothesis implies, the relation between vertical radial versus horizontal tangential movements is the main factor in the L-illusion, the illusion should differ significantly between ipsilateral and contralateral scanning, but not between the hands being used for scanning. Some laterality effects, albeit in some complex interactions, had been found by Heller, Joyner, and Dan-Fodio (1993). Movement effects were, therefore, examined in relation to laterality and left and right hemispace (Millar & Al-Attar, 2000).

Twenty right-handed (by self-report) blindfolded student volunteers were randomly assigned to using the forefinger of the right hand or the left hand to explore L-shapes in tabletop space. Highly tangible L-shapes were used. The shapes were embossed professionally on transparent plastic sheets. The L-shapes consisted of an 8-cm vertical standard and one of five horizontal comparison lengths (7.0, 7.5, 8.0, 8.5, 9.0 cm). Each test sheet contained two shapes, which started and ended, respectively, in right and left hemispace at 11 cm from the middle of the sheet.

The task was to scan the vertical standard and to adjust the horizontal comparison line to the length judged equal to the standard. The experimenter placed the participant's finger at the start of the standard line in every trial. The participant was to move the finger down to the comparison line and to scan it up to, or beyond, the end of that line to whatever length she or he judged to be equal to the standard length. The starting location was counterbalanced over trials.

Overshooting the comparison line by a length that exceeded the vertical standard was scored as a positive error. Undershooting errors, showing that the vertical standard was underestimated, were scored as negative. The mean constant (signed) errors show the point of subjective equality (PSE) that indicates the presence or absence of a positive or negative illusion for each hand separately (Figures 6.2a & 6.2b).

The important results were the average constant (signed) errors that show the PSE that indicates the presence or absence of a positive or negative illusion for each hand separately in ipsilateral and contralateral hemispace. The PSE for the right hand showed a small but significant negative illusion that was the same in both the right and left hemispace. The left hand produced no illusion. The PSE for the left hand did not differ significantly from zero in either ipsilateral and contralateral space.

The hypothesis that the difference between radial and tangential movements is the crucial factor in the haptic vertical–horizontal illusion was not supported. Reaching across to the contralateral side would involve tangential shoulder movements more than scanning in ipsilateral space, without necessarily affecting radial arm movements. But the left hand produced no illusion in either hemispace. Moreover, the small negative illusion for the right hand was also the same in left and right hemispace.

The highly significant effect that is obvious from the graphs was due to the set of comparison lengths. As in other studies with psychophysical comparison methods (Armstrong & Marks, 1997, 1999; Gescheider, 1997), the lengths of comparison stimuli have considerable effects. The constant errors

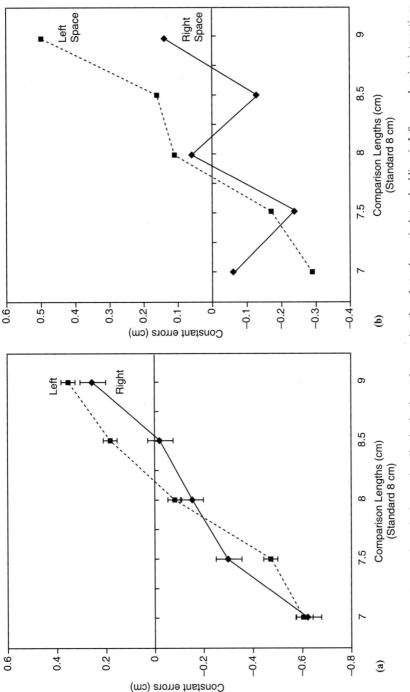

Figure 6.2 Mean constant and standard errors in adjusting horizontal comparison lengths to the vertical standard lines in L figures, showing interactions of the range of five stimulus lengths (standard 8 cm) in right hemispace (continuous line) and left hemispace (dotted line) for the right hand (**a**) and the left hand (**b**). [From S. Millar & Z. Al-Attar, "Vertical and bisection bias in active touch", *Perception*, 2000, 29, 481–500. Reproduced with permission from the publisher, Pion Ltd., London.]

here varied from negative to positive for lengths that were, respectively, shorter and longer than the standard in both left and right hemispace for the right hand, and in left hemispace also for the left hand. A significant interaction of hands with comparison lengths and hemispaces showed that for the left hand the comparison length effect did not reach significance in the right hemispace. It was not clear why. It may have involved a small left-hand/right hemisphere spatial advantage that also eliminated any vertical–horizontal bias completely (e.g. Heller, Joyner, & Dan-Fodio, 1993). But whatever the explanation, the zero illusion was the same for the left hand in contralateral as well as ipsilateral hemispace.

The small but significant negative illusion shown by the PSE for the right hand could have been produced by using two L-shapes in the layout. The vertical lines of L shapes, scanned in counterbalanced successive trials, were parallel to each other, albeit at a distance. It is possible that this reduced their length estimation below that of the horizontal comparison line. If so, it would have had less effect on left-hand scanning if that indeed involved more accurate spatial coding. We consequently provided spatial cues for the vertical lines in the next experiment by aligning the vertical lines of L-shapes to the body-midline.

Curiously enough, few researchers seem to have applied the Künnapas hypothesis for vision to vertical overestimations in haptic perception. As noted earlier, Künnapas proposed that the main factor, albeit not the only one, in the visual illusion was the spatial relation of the perceived lengths to the boundaries of the visual field. In terms of reference information this means that the perceived length depends on the relation of the stimulus length to the available egocentric visual reference cues. The reference hypothesis assumes that, in the absence of vision, the relevant egocentric reference cues for inputs from touch and movement would be centred on body and posture cues. But the same rule should apply.

The prediction was, therefore, that L-shapes in which the vertical line is aligned to the (sagittal) body midline will not produce either positive or negative vertical–horizontal illusions, even when the shape is explored with the right hand rather than with the left hand.

It was also important to test the hypothesis that the haptic vertical–horizontal illusion differs from length illusions produced by shapes that contain bisecting lines. Both L-shapes and inverted T-shapes were therefore tested in a second experiment.

The main conditions and procedures were the same as before, except that only one stimulus shape was presented in any trial. Both the L- and T-shapes were 8-cm vertical standards, and they were judged by adjusting one of five (from 7.0 to 9.0 cm) horizontal comparison lines to the standard, as before. The vertical standard in both L- and inverted T-shapes was aligned to the participant's body midline. Twenty right-handed, naive volunteers were

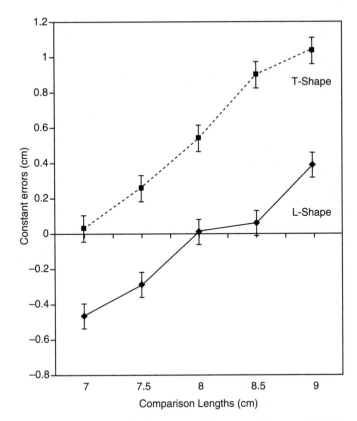

Figure 6.3 Mean constant errors for L-shapes and for inverted T-shapes ("A"), showing the PSE when the vertical lines in each were aligned to the body midline. [From S. Millar & Z. Al-Attar, "Vertical and bisection bias in active touch", *Perception*, 2000, *29*, 481–500. Reproduced with permission from the publisher, Pion Ltd., London.]

allocated randomly to the two shapes. They were blindfolded and told to use only the forefinger of the right hand. To ensure that participants explored the whole of the inverted T-shape, they were told to move the finger down and then to the left and then right to ensure that the whole of the horizontal comparison line was scanned before adjusting its length relative to the vertical standard. The results showed a highly significant difference between L- and inverted T-shapes (Figure 6.3).

As predicted, the L-shape produced no illusion when the vertical shaft was aligned to the body-centred (sagittal plane) midline. The PSE for the L-shape did not differ significantly from zero, although the right hand was used for scanning. The finding suggests that spatial reference information was involved, as indeed is implied by previous studies that showed varying results for the haptic vertical–horizontal illusion in different spatial planes.

Providing body-centred reference information for L-shapes by aligning the vertical shaft to the body midline in the sagittal plane cues thus produced accurate judgement of the horizontal length relative to the vertical shaft.

By contrast, the PSE for inverted T-shapes was positive and highly significant. Thus, although the vertical shaft of the inverted T-shape had also been aligned to the body midline, the PSE for the inverted T-shapes showed a highly significant overestimation of the vertical compared to the horizontal line. It implies that the illusion differed for the two shapes, suggesting that the two illusions did not depend on the same factor.

The comparison length effect was highly significant for both shapes, as shown by the error slopes, as in other psychophysical length judgements. However, both underestimations and overestimations were proportionate to the main PSE effect, respectively, for L-shapes and for inverted T-shapes. All errors were positive for inverted T-shapes, but proportionately negative and positive for L-shapes. The significant slope effect was thus independent of the difference in PSE for the two shapes. It cannot, therefore, be considered a contributing factor to the presence or absence of the illusion for either shape here.

The findings show that the factors that produced the overestimation of the vertical shaft in inverted T-shapes differ from the factors that produce or eliminate the vertical–horizontal illusion. The vertical shafts of both configurations were aligned to the body midline. That did not prevent the very significant T-shape illusion.

By contrast, the findings on L-shapes in the two experiments suggest that body-centred spatial information is involved in the vertical–horizontal illusion. Aligning the vertical shaft of the L-shape to the body midline eliminated the negative L-shape illusion found for the right hand in the previous experiment, as predicted.

The reference hypothesis can, in principle, explain the findings from previous studies and from the present findings on the haptic vertical–horizontal illusion. The assumption is that the haptic vertical–horizontal illusion, when it occurs, depends, as does the visual vertical–horizontal illusion (Künnapas, 1957b, 1957c), on discrepancies between compressed (elliptical in the case of vision) reference spaces and on inputs that imply a more extended spatial field.

Haptic reference information is based on relating arm and hand and finger movements to cues from body postures. The notional compression of space for haptic inputs must therefore depend on discrepancies between body- and posture-based spatial reference cues for different arm, hand and finger movements. Horizontal "tangential" movements, anchored at the shoulder and sweeping around the body, describe a large arc around the body. The length of a given horizontal movement thus takes up a relatively small proportion of the spatial "field" with respect to the shoulder as the anchor point. By contrast, the total possible extent of a vertical or radial movement that is centred on the elbow is relatively small.. A given "vertical" movement, anchored at the elbow joint, is thus likely to constitute a much larger proportion of the possible extent or area covered by the movement. The haptic vertical–horizontal illusion, when it occurs, can thus be explained by a similar, vertically compressed and horizontally extended reference area as the visual vertical–horizontal illusion.

The presence or absence of the vertical–horizontal L-illusion in touch, as in vision, can thus be explained by the presence or absence of discrepancies between inputs in relation to the available reference information in different spatial planes.

The very variability of vertical–horizontal illusions for L-shapes in different studies makes good sense when considered in terms not of movements as such, but of movements in relation to different body-centred cues that provide reference points for the extent of arm, hand and finger movements. Arm movements that are anchored to the shoulder as the reference point produce longer movements, and they scan a potentially larger area than movements that are anchored to the elbow joint as the point of reference. The radial/tangential movement hypothesis is thus substantially correct. But it does not go far enough. It leaves out of account that judging the extent of arm movements also depends on the spatial plane in which the movements are executed and on reference cues from the shoulder, elbow joint and wrist or finger joint that provide anchor or reference points for given length judgements.

Interestingly enough, findings with L-shapes in passive touch (Wong, Ho, & Ho, 1974) are fully consonant with that principle. When they placed an L-shape on the forearm with the (normally) vertical component aligned along the longitudinal axis of the forearm, so that the (normally) horizontal component line lay across the forearm, the horizontal component was judged longer than the component aligned with the longitudinal axis of the forearm. The authors concluded that the vertical–horizontal (or horizontal–vertical) illusion in passive touch is a function of the orientation of the components with reference to the elongated frame of the receptor organ. They assume that this effect is independent of the effect in vision and in active touch. In fact, however, the rule that length perception is influenced by the spatial context in which it occurs or to which it is related applies both to vision and to active touch.

The same explanation is not sufficient for the very substantial T-shape illusion in the experiment just considered. It occurred in exactly the same tabletop space, with the same standard and comparison lengths and the same body alignments as for L-shapes. Hatwell (1960) found the bisection illusion, but no vertical–horizontal illusion, in touch in tabletop space. The difference is supported by the results here.

However, there had been no previous attempt to explain how or why shapes that include line bisections produce illusions. We therefore ran a series of studies to explore bisection and sectioning effects further with haptic shapes.

ILLUSIONS IN SHAPES CONSISTING OF BISECTING AND BISECTED LINES

Here again it was worth applying the analysis by Künnapas (1955a, 1955b) of the visual illusion to haptic T-shapes.

Künnapas showed clearly that the visual bisection illusion does not depend on the overestimation of vertical lines, as such, but is due to bisection effects. Vertical lines in T-shapes are overestimated if they bisect horizontal lines, but not when they are bisected by horizontal lines.

In terms of categorising spatial reference information into the three broad categories of external (allocentric), body-centred (egocentric) and object- or shape-based reference information, bisection effects – if and when they occur – belong to the third category, that of shape-based reference information. The bisecting and the bisected line, which produce the illusion, are features of the same shape.

The first thing was to test whether haptic bisection effects produce an illusion regardless of the orientation of the vertical line. We also assessed a possible short-term memory effect. The vertical bisecting lines in inverted T shapes had been previously been scanned first. Results for the shapes ("A", "B", "C", "D" in Figure 6.4) were therefore compared. Results for the inverted T-shape "A", in which the vertical line was scanned first, were compared with inverted T-shapes in which the vertical bisecting line was scanned after the bisected horizontal ("B"). T-shapes rotated 90 degrees to the right ("C"), and T-shapes rotated 90 degrees to the left ("D") were used to test whether bisection effects are due to the vertical orientation or to the bisection of one line by another. For shapes labelled "C", the horizontal was the bisecting line that was scanned before the vertical bisected line. For shapes labelled "D", the vertical was the bisected line that was scanned before the horizontal bisecting line.

High school student volunteers were randomly allocated to the four shape conditions and were tested blindfold. The judgement task, adjusting comparison lengths, scoring in terms of vertical lines, and all other procedures were otherwise the same as before. The points of subjective equality and standard errors for the four types of shape, graphed in Figure 6.4, show that vertical lines are judged to be longer than horizontal lines because they are the bisecting lines in inverted T-shapes, and not because they are oriented vertically. The bias was reversed in rotated T-shapes in which the horizontal line was the bisecting line, and the vertical line was bisected.

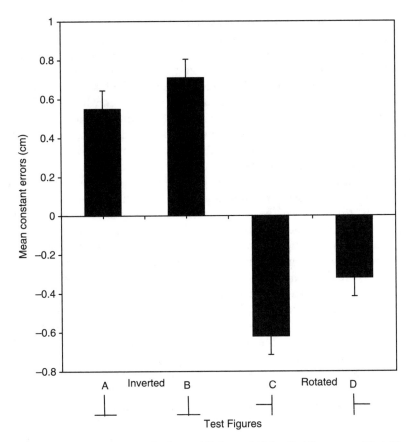

Figure 6.4 Mean constant errors for inverted T-shapes "A" (vertical line, scanned first, bisects the horizontal line), "B" (inverted T-shape, vertical line scanned second, bisects horizontal line), "C" (rotated shape: vertical line, scanned second, is bisected by the horizontal line), and "D" (rotated shape: vertical line, scanned first, is bisected by the horizontal line). [From S. Millar & Z. Al-Attar, "Vertical and bisection bias in active touch", *Perception*, 2000, *29*, 481–500. Reproduced with permission from the publisher, Pion Ltd., London.]

The bisecting vertical line was very significantly overestimated in both the inverted T-shapes "A" and "B". By contrast, the bisected vertical line in rotated shapes was significantly underestimated, showing a significant negative PSE for both the "C" and "D" shapes.

The findings show that the bisection illusion is due to bisecting versus bisected lines, and not to the vertical or horizontal orientation of lines in these shapes. Calculating the adjustments, as before, in relation to vertical lines, the difference between inverted T-shapes in which verticals bisect horizontals, and rotated T-shapes in which horizontals bisect the verticals was highly significant.

A possible additional memory effect could not be ruled out. Scanning the bisected horizontal line before scanning the bisecting vertical line ("B") increased the positive overestimation of the bisecting vertical line compared to scanning the bisecting vertical line first ("A"), albeit not significantly. Scanning the bisected vertical line before scanning the bisecting horizontal line ("D") decreased its negative effect relative to ("C"), in which the bisected vertical line was scanned after scanning the bisecting horizontal line. We originally (Millar & Al-Attar, 2000) attributed these differences to the decay of movement information in immediate memory (Pepper & Herman, 1970). It could explain why adding positive and negative effects left a small, positive residue, although adding percentages of positive and negative effects is not strictly justified with separate subject groups. Nevertheless, we ran a control experiment to assess whether the "residue" could or should be attributed to an additional vertical–horizontal (radial/tangential) effect, as is often assumed on the ground that radial (vertical) movements are slower than tangential horizontal movements. However, the interpretation that radial movements are slower had only been tested relatively indirectly (Wong, 1977).

The control experiment here used the video-recording and timing (1/100-s) device described in Chapter 4. Inverted T-shapes ("A") were produced on transparent plastic sheets, as before, and presented on the transparent reading surface of the device. Ten new volunteers were tested blindfolded, using the same task and procedures as before.

As expected, the PSE was significantly positive, confirming the illusion. However, the latencies for vertical and horizontal (90 degree) scanning differed according to whether the movement time was taken from the start to the end of the left–right horizontal length, or whether it included the right–left horizontal backtracking movement that was necessary initially to get to the start of the horizontal line from the bisection point of the comparison length. Only when the horizontal backtracking movement was excluded were the latencies for the vertical line longer than for the horizontal line. At the same time, neither the correlations of latencies with constant errors, nor a simple regression analysis (z-scores for vertical errors, keeping horizontal latencies constant), were significant.

More important, horizontal (tangential movement) latencies that included the actual horizontal (tangential) backtracking movements were very significantly slower than the latencies for the vertical line. The vertical line was overestimated nevertheless. The results showed clearly that simple differences in latencies for actual vertical and horizontal movements do not explain the perceptual bias. It must be assumed at least that participants actually discount horizontal backtracking movements.

Simple left–right scanning could well be faster simply as the more familiar movement in writing. Moreover, slower movements are often overestimated. But the fact that the total tangential scanning movement actually took longer than the vertical movement without affecting, let alone reversing, the positive illusion further suggests that movement-time differences would not explain why positive and negative bisection did not simply cancel out. A small short-term memory effect due to scanning the constituent lines first or second could not be ruled out. But that would not account for the highly significant difference in positive versus negative PSE errors between inverted and rotated T-shapes.

The control experiment thus produced nothing that altered the very significant main finding that the main factor in the T-illusion is the difference between bisecting and bisected lines within the shapes, and not the orientation of the lines in the shapes.

Having found the effect in touch that Künnapas showed in vision, the important point was to explain why the presence of a bisecting and a bisected line should produce length illusions. As noted earlier, there have been few or no previous attempts to explain bisection effects beyond reporting that they occur.

It was argued earlier that all types of T-shapes consist of a bisecting and a bisected line. The bias thus results from discrepant length information from within the shapes. The clear difference in perceived length of the bisecting compared to the bisected line strongly suggests that the illusion is due to discrepant length information from the two lines. Objectively, the two lines are equal in length. Moreover, when separate vertical and horizontal lines of the same length are compared, there is no illusion (Heller et al., 1997). The intersection is crucial.

It is assumed here that the perceived difference in length when one line bisects the other results from a difference in reference, or anchor, cues between the two lines. The continuous bisecting line, whether vertical or horizontal, is bounded only by its start and end points. The bisected line is also bounded by a start and an end point. But the bisecting line creates a further boundary in the middle of the bisected line at the point where it bisects that line. The point that divides the bisected line is felt as a boundary point between the two halves of the bisected line. It thus adds a further potential reference point, which halves the bisected line (Newcombe & Liben, 1982).

The junction point thus produces a disparity in boundary or anchor points between the continuous bisecting line that is only limited by the start and end points of the line or movement and the bisected line that also has a further boundary or mid-anchor point. Each section of the bisected line is only half the length of the continuous bisecting line. Seeing the smaller line on each side of the midpoint of the bisected line biases the length judgement of the continuous line, making it look longer. Similarly, in touch, the short movement on either side of the T-shape biases the length judgement of the continuous bisecting line, which thus feels longer.

The length of the movement that precedes the length judgement or adjustment is thus likely to influence that adjustment. The small extent of a preceding vertical or horizontal section beyond the junction would affect the judgement of the whole of the bisecting line and therefore lead to an overestimation of the continuous line, regardless of its orientation.

We argued that a test of whether the junction point really does bias the length judgement of the continuous line, which does the sectioning, is to vary the sectioning point of the continuous line by moving it along the sectioned line. If the hypothesis is correct, the smaller the movement length beyond the sectioning point becomes, the larger would be its effect on judging the length of the continuous sectioning movement. The hypothesis thus assumes that the shorter the movement beyond the junction of a bisected line becomes, the greater will be the overestimation of the continuous bisecting line.

In order to test the prediction, we used five "junction" shapes, embossed as before. In each shape the vertical (8-cm) continuous line joins the (8-cm) horizontal comparison line at one of five different locations. The junctions were respectively at 0 cm distance from the start (J0 = L-shape), and at 2 cm (J2), at 4 cm (J4), at 6 cm (J6) and at 8cm (J8 = reversed L-shape).

Neither the L-shape (J0), nor the reversed L-shape (J8) were strictly relevant. The vertical standard line does not section the horizontal comparison line in either case. However, the L-shape and the inverted L-shapes were included for interest. Little or no effect was expected for the L-shape, because it does not section the horizontal line. Any comparison judgement would depend on the whole length of the horizontal line. No prediction was possible for the reversed L-shape (J8). Not only does the shape not involve any intersection, no horizontal movement beyond the vertical line was expected that could influence the judgement. It was included, nevertheless, to see how that shape would be treated.

The main prediction for the other sectioned shapes was that the illusion would increase inversely with the length of movement beyond the sectioning or reference point. The shortest movement beyond the sectioning or reference point would thus bias length judgements most and produce the largest illusion.

Thirty-one volunteers were tested blindfolded on all shapes. The order of presentation was counterbalanced over runs and participants. The procedures were otherwise the same as for inverted T-shapes ("A"). The shapes and results are shown in Figure 6.5.

The hypothesis received very significant support. The location of the junction point on the sectioned line was highly significant. It showed that the degree of overestimation of the vertical line varied with the length of movement beyond the junction point: the shorter the movement, the larger the overestimation.

As predicted, the overestimation bias increased with decreasing lengths of the sections beyond the junction point. The largest PSE was for junction J6

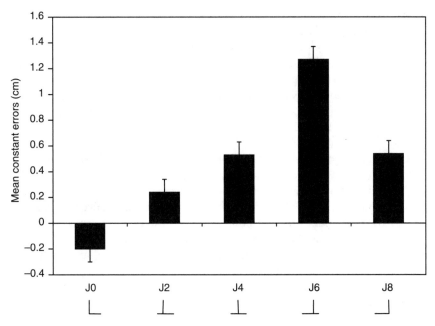

Figure 6.5 Junction figures: Mean constant errors and error bars in adjusting horizontal (8-cm) standards to vertical (8-cm) standards in shapes with junctions that bisect the horizontal line at J0 (0 cm), J2 (2 cm), J4 (4 cm), J6 (6 cm) and J8 (8 cm). [From S. Millar & Z. Al-Attar, "Vertical and bisection bias in active touch", *Perception*, 2000, *29*, 481–500. Reproduced with permission from the publisher, Pion Ltd., London.]

(15.88%), with the smallest post-junction projection. Neither the L-shape nor the reversed L-shape have any intersection points. The results are thus not strictly relevant to the hypothesis, but are of interest nevertheless. The L-shape (J0) showed a slight negative PSE. The reversed L-shape (L8) was not sectioned by the vertical line either, and it had no objective protruding length that extended beyond the vertical line. However, most participants actually produced overshoots of the horizontal line beyond the junction with the vertical. Overshooting the horizontal in the reversed L-shape should probably have been expected. It was the only one of the five stimuli in the set in which the horizontal line did not protrude beyond the junction with the vertical line. A context effect of that kind should have been anticipated, although the order of the stimuli was counterbalanced. The overshooting error also explained why the overestimation bias for J8 was only less extreme than that for J6, which produced the largest bias and had the smallest length protruding the junction point.

The results were thus consistent with the prediction that the junction point did "section" the bisected line into two smaller lengths, and that the shorter length or scanning movement, beyond the junction point, influenced the comparison with the undivided sectioning line, making it seem much larger. The findings imply that bisection effects are due to the junction point on the sectioned line. The junction point is perceived as a stop or reference border between two smaller lengths. The smaller length beyond it affects the comparison judgement. Length judgements for the comparison lengths are thus influenced by stop points that delimit the constituent length inputs. The additional boundary point on the bisected line thus biases the length estimation of the undivided bisecting line, so that the continuous line is judged to be larger.

A further question was whether, or how, the bisection illusion relates to external and body-centred reference cues (Millar & Al-Attar, 2000, Exp. 6). The reference hypothesis implies that, in principle, additional congruent external and body-centred reference information should reduce the bias created by the discrepant junction points between the two lines that constitute T-type shapes. The additional reference information should override, or at least diminish, the discrepancies in reference cues for length that are inherent in these shapes. Ameliorating effects of body-centred reference information were found, for instance, in the perception of shape symmetry, which is much less effective in the perception of tactual shapes that are explored with one finger than it is in vision (Ballesteros, Millar, & Reales, 1998; Millar, 1978).

Inverted T-shapes (Type "A") were prepared on transparent sheets as before and were fastened to rigid square plastic plates. The square edges of each plate provided a highly tangible square frame for each stimulus shape. Twenty new high school volunteers were randomly allocated to the reference and frame instructions conditions. The control conditions were the same as for the usual T-shapes (Type "A") in which the bisecting line is vertical (in tabletop space) and aligned to the body midline. A sizeable bisection illusion could, therefore, be expected in the control conditions. But the "frame" instructions should reduce them.

Procedures for participants in the reference instruction condition were thus the same as for T-shapes (Type "A). But they were also instructed to use their other hand to explore the frame that surrounded the shape and to relate the external frame to the shape. They were also told that the vertical shaft of the shape, as projected in tabletop space, was aligned to their body midline. They were to relate the inverted T-shape also to that body-midline cue.

The set of comparison lengths had the expected highly significant effect, as before, showing again the importance of context effects from the set of stimulus lengths being used. As before, the context effect was independent of the illusion. Different lengths decreased, or increased, the overestimation of the bisecting line proportionately to the PSE.

The main results showed that the reference instructions reduced the illusion very significantly, as predicted. But although the use of reference cues more than halved the illusion, from 6.88% to 3%, the combined reference instructions did not eliminate the illusion.

Perceptual illusions have practical implications when they are produced in maps, or other spatial displays, that provide symbolic information about large-scale space. Perceptual illusions occur in visual maps (Gillan, Schmidt, & Hanowski, 1999). We assumed that they would also show up in tactual maps (Millar & Al-Attar, 2001).

We used a raised-line route in a map that included an inverted T-junction in a tactual map with naïve right-handed blindfolded sighted volunteers. Their task was to indicate a point on the raised-line route at which the second part of the route (bisecting or bisected lines as counterbalanced) equalled the first (bisected or bisecting part of the route, as appropriate). The bisection error for the inverted T-junction was highly significant (11.83%). The overestimation of the continuous part of the route was also highly significant (11.33%) when a landmark was felt near one section of the bisected route.

We again used dual (body-centred and external) reference instructions in a further condition to test effects on the illusion. The dual instructions consisted of telling participants to use their body midline as a reference cue and also to scan a tangible rectangular frame that surrounded the map with their free (left) hand. They were to relate the frame information and their body midline to the comparison route when judging the line lengths. They were to indicate the point at which the comparison (bisected line) length equalled the initial part of the route.

The result was quite startling. The reference instructions reduced the overestimation of the continuous sectioning line to zero.

The assumption had been that using both body-centred and external reference instructions would have greater effects in overriding or ameliorating the bisection illusion than using only one form of reference information. That assumption may have been mistaken with respect to discrepant reference cues, as shown in the previous study (Millar & Al-Attar, 2000), since the bisection illusion was not eliminated, although it was more than halved. With hindsight, it seems likely that the instruction to use external cues as well as body-centred cues for reference was a mistake. Bisection illusions are known to occur in vision quite as strongly as in touch, despite the fact that external cues are present in vision. The total elimination of the tactual bisection illusion in the map study (Millar & Al-Attar, 2001), compared to the earlier finding, thus suggested that the realistic map context may have had an additional effect.

CONCLUSIONS

This chapter has examined the question of whether it is simply by chance that "visual" illusions also occur in touch. The Helmholtz principle that perceptual illusions are due to discrepancies in the very factors that produce accurate perception when they converge or coincide seemed to me to suggest otherwise. The reference hypothesis assumes that accurate shape and spatial perception depends on convergent reference cues. It implies that discrepancies in the potentially available reference cues can produce illusory biases. The vertical–horizontal illusion was of particular interest in that regard. The shapes that produce it in vision and in touch consist of adjacent vertical and horizontal lines of equal length, in which the vertical line appears to be longer than the horizontal line. The illusion had often been studied with L-shaped and with inverted T-shaped figures. But the illusion is not the same for the two shapes in either vision or touch. The chapter discussed findings that suggest that the explanations for the two types of illusion also differ, but that they apply to both the visual and the haptic illusions.

As the name itself implies, the vertical–horizontal (or horizontal–vertical) illusion is orientation-specific. The visual illusion was typically explained as an optic phenomenon. But a series of previous studies had shown that the overestimation of the vertical line related to the configuration of the potential visual field in which vertical compared to horizontal lines were seen. The L-shape haptic illusion had previously been explained by differences in the arm movements required for horizontal and vertical movements in scanning L-shapes. Horizontal (tangential) arm movements are centred on the shoulder joint. Vertical (radial) arm movements are centred on the elbow joint for the vertical line. The difference was assumed to relate to differences in movement speed. However, the haptic L-shape illusion had previously been found only in some, but not all, spatial planes. Vertical overestimations were not found in the fronto-parallel plane and do not seem to occur in tabletop space. The further studies considered here showed no overestimation of the vertical line in small L-shapes in tabletop space, consistent with previous findings, and despite the fact that the effect was tested with movements that were thought to involve the shoulder joint more by reaching into contralateral reaching, compared to reaching to the same side. Moreover, even the small negative illusion was eliminated by aligning the vertical shaft of L-shapes to the body midline in tabletop space.

The spatial plane in which the haptic vertical–horizontal illusion is tested is thus evidently relevant in considering the conditions in which the vertical–horizontal illusion occurs in touch. The presence of a haptic L-illusion may well be due to an overestimation of radial movements anchored at the elbow joint compared to tangential movements that are anchored to the shoulder, and this may well be correct. But it is not sufficient. It leaves out of account that horizontal movements for L-shapes constitute a relatively small proportion of the total possible extent of tangential movements around the body. The movement for the same vertical length is a much larger proportion of the total sweep than is possible for radial movements that are anchored at the elbow. It would thus occupy a much larger proportion of the area or possible movement extent relative to the body and would consequently be overestimated. The rule is the same as that proposed by Künnapas for the visual illusion relative to an elliptical visual field. It implies a common factor or common spatial rule for the vertical–horizontal illusion in vision, and for its occurrence in active touch or in passive touch. The interpretation needs to be checked out further. But it is a more likely explanation of the similarity of haptic vertical–horizontal illusions, when they occur, than the assumption that the similarity is a chance effect or that haptic perception shows it less because touch is more veridical.

The T-shape illusion differs. It is not orientation-specific in vision. The empirical findings that were considered here also show that it is not orientation-specific in touch either. The T-shape illusions occurred in touch, regardless of the vertical orientation of the bisecting line. The new findings that were considered in this chapter showed that tactual illusion, like the visual T-shape illusion, is due to the fact that one of the constituent lines of the shape bisects the other. The findings supported the further hypothesis that the illusion is due to an imbalance of length cues within the shape. The point of bisection on the bisected line was shown to determine the overestimation of the continuous bisecting line. The smaller the line protruding beyond the section point, the larger was the overestimation of the continuous line that had no such additional reference point for estimating its length.

The findings for the bisection illusion supported the explanation that junction points add additional reference cues for length judgements that compare the sectioned line to the sectioning line. Shifting the sectioning point along the sectioned line had significant and consistent effects on the size of the illusion. The bisecting line produces an additional boundary point on the sectioned line, which divides the sectioned line into smaller portions. These produce the overestimation of the continuous bisecting line. The visual bisection illusion is also due to the overestimation of the continuous bisecting line. The explanation is that the bisecting point acts as a boundary between the two much smaller sections of the bisected line. It is assumed that these smaller extents bias perception. That should also apply in vision.

The findings suggested that the similarities to vision of different illusions are not chance effects. They can be explained by discrepancies in normally consonant reference information from either external, body-centred and/or or shape-based sources.

The discrepancies in reference cues that produce the vertical–horizontal illusion and the bisection illusion differ. The vertical–horizontal illusion occurs by length inputs that relate to external and/or body-centred spatial reference information.

The bisection or T-type illusion, by contrast, was explained here as a mainly shape-based illusion. However, the relation of the haptic bisection illusion to the combined body-centred and external reference information raised a further question. The bias in inverted T-shapes was significantly reduced by concomitant body-centred and external reference instructions, consistent with the assumption that the dual instruction would help to override the biasing effects of bisection points. But the bisection bias was still significant. It suggested that the illusion is mainly due to the discrepancy in length cues within the shape. But the remaining illusion also suggested that using both forms of reference in combination, instead of doubling the effect of overriding discrepant reference cues from within the shape, as had been assumed, was evidently not completely effective.

The findings contrasted with the result of the further study, which used a "T-junction" in a tactile map. The section illusion was also highly significant in the tactual map. But the same combination of reference instructions actually eliminated the quite strong illusion completely in the map context. The map context seems to have had an additional effect in reducing the illusion.

The relation of the shape-based bisection illusion to external and body-centred reference information is considered further in the next chapter.

CHAPTER SEVEN

Müller-Lyer shapes

Müller-Lyer shapes produce powerful involuntary illusions from which it is difficult to escape. The shapes consist of a shaft that is flanked at each end by angled wings or fins that either diverge outwards from the shaft or are angled inwards towards the shaft. The illusion consists of seeing the shaft with diverging fins as much longer than the shaft in shapes with converging fins.

The Müller-Lyer illusion presents us with a nice puzzle that challenges the time-honoured division into "top-down" and "bottom-up" theories of perception perhaps more than any other perceptual effect. The Müller-Lyer illusion does not diminish when you know that the size of shaft in the figure you see is illusory. It persists even if you measure the shafts of shapes with diverging and converging fins. Higher-order knowledge cannot, therefore, explain the visual illusion. The illusion seems to be due to low-level visual mechanisms that are not affected by higher-order knowledge (e.g. Ebert & Pollack, 1972). On the other hand, low-level visual factors cannot explain why the visual Müller-Lyer illusion occurs also in touch.

The intrinsic interest of the illusion and its relevance to the subject matter of this book must serve as the excuse for devoting a whole chapter to it. Previous findings and studies will be discussed first. Two intriguing hypotheses emerge from it. They suggest possible factors that the visual and haptic illusions may have in common. The possibilities are checked out in experiments that are described in some detail next.

The Helmholtz (1867, 1896) principle that illusions are due to discrepancies in the very cues that normally produce accurate perception was also applied to the Müller-Lyer illusion. It is argued that the illusion in both touch and vision is due to discrepant length and size cues from the features that constitute the shapes.

The hypothesis is supported by findings that show a very significant haptic illusion for both horizontally and for vertically oriented Müller-Lyer shapes. Further findings also show that external reference information, far from overriding the illusion (DeLucia, Longmire, & Kennish, 1994; Gentilucci, Chieffi, Daprati, Saetti, & Toni, 1996), actually increases it. By contrast, instructions to use body-centred cues for reference in adjusting the comparison shape reduced the Müller-Lyer illusion to a near zero level in touch and also in vision.

The implications of the findings on the Müller-Lyer illusion and for the vertical–horizontal and bisection illusions described in the previous chapter are compared and discussed in the final section. Considered together, the findings suggest that modality-specific and contextual factors contribute to all three of the illusions. But the main factor in the three illusions is the discrepancies in different types of relational information that normally produce accurate shape perception in touch as well as in vision. In the vertical–horizontal illusion the discrepancy is produced by the relation of the vertical and horizontal length stimuli to the boundaries of their respective perceptual "fields". The visual "field" is more compressed vertically than horizontally. Vertical length stimuli thus look larger with respect to the smaller-possible vertical extent. Horizontal lengths look smaller with respect to the wider horizontal extent of the elliptical field, and thus look smaller in relation to the same vertical lengths. The same relative discrepancy applies in haptics, although people are not aware of it in either modality. Vertical lengths are produced by radial movements that are anchored at the elbow joint. Horizontal lengths are produced by tangential movements that are anchored at the shoulder joint. The total possible extent or area of radial movements is small compared to the total possible extent or area of tangential movements that are anchored at the shoulder joint. Vertical length stimuli that depend on radial movements are thus perceived as relatively large with respect to the small total extent or spatial boundary of radial movements, and as larger than the same horizontal length stimuli that depend on tangential movements that relate to the larger extent or spatial boundary. By contrast, the previously discussed bisection illusion, and the Müller-Lyer illusion that is considered here, are shape-based illusions. The shapes consist of constituent features that provide discrepant length and size cues about the configuration. The conclusions are discussed in the final section.

PREVIOUS EVIDENCE ON THE MÜLLER-LYER ILLUSION

Over (1968) presented a review that covered almost all the explanations of the Müller-Lyer illusion put forward then or since. He considered none of them satisfactory. According to him, low-level visual accounts do not explain the haptic geometric illusion. But theories that attribute the illusions to higher-order perceptual strategies have rarely been specified sufficiently to allow direct tests.

A somewhat problematic remedy for the evident contradictions in both "low-level" visual explanations and "higher-order knowledge" accounts is simply to reject one of them. Thus, the Müller-Lyer illusion is still often considered a purely visual phenomenon (Walker, 1971). The fact that it occurs in touch is attributed to attempts to imagine the shapes visually.

It has been shown that vision can affect the haptic illusion (Walker, 1971). The illusion was increased when people could see, but could not touch, the fins of a Müller-Lyer block-shape when they were palpating the shaft of the block-shape without seeing it. Sighted people often report visual imagery, or attempts to see the shape in their "mind's eye", when asked to judge Müller-Lyer shapes by touch. A study with students of psychology found significant (albeit not very high) correlations between the strength of the illusion, imagery reports and scores on an imagery questionnaire test (Frisby & Davies, 1971). However, that probably depends on individual differences. Heller and his colleagues (2002) tested blindfolded-sighted people, people with low vision, and late blind as well as congenitally blind people, and found no effect of visual status on the strength of the Müller-Lyer illusion. Shapes with acute angles produced the strongest illusion.

The fact that the illusion also occurs in touch has been known for almost as long as its discovery in vision (Fry, 1975; Judd, 1899, 1902, 1905; Pearce, 1904; Rieber, 1903; Robertson, 1902). Moreover, the illusion has been found as much in congenitally blind people as in the sighted (Bean, 1938; Casla, Blanco, & Travieso, 1999; Heller et al., 2002; Patterson & Deffenbacher, 1972; Révész, 1934; Robertson, 1902; Tsai, 1967). As noted earlier, the use of the term "blindness", or even "early blindness", does not always exclude participants who have had some visual experience. However, there is evidence that the illusion is shown by people who are definitely known to have been totally blind from birth (Heller, personal communication, 2005; see also Heller et al., 2002).

Over (1968) argued against an explanation of the illusion in terms of inappropriate size-constancy scaling (Gregory, 1963). The hypothesis assumes that the processes that make objects look the same size at different viewing distances in 3-D space also produce illusory 3-D depth cues in 2-D pictorial space and consequent geometric illusions.

There are some intriguing links between the detection of pictorial depth cues and sensitivity to geometric illusions. A man who had been blind from the age of 10 months, and regained sight through cataract operations relatively late in life, recognised simple shapes and objects by sight alone. But he was insensitive to pictorial depth cues and to common geometric illusions (Gregory & Wallace, 1963). Another example is that of a man who was blinded at the age of 2 years and showed good visual object recognition after cataract operations in mid-life. But he failed to detect pictorial depth cues in 2-D drawings. He also failed to show the depth-reversal illusions that 2-D drawings of cubes and staircase figures normally produce (Fine et al., 2003). Whether or how experience with binocular depth cues relates to detecting depth cues in 2-D pictures is not yet clear. More recently, a patient with acquired visual agnosia was found to be severely impaired in extracting 3-D information from 2-D pictures and also failed to show the Müller-Lyer and Ponzo illusions (Turnbull, Driver, & McCarthy, 2004).

Gregory's (1966, 1990, 1998, 2003) explanation is persuasive because it covers a number of different visual illusions. But there are also many findings that are inconsistent with some implications of the inappropriate size-constancy hypothesis. That probably means that translating 3-D rules to 2-D shapes contributes to visual illusions, but is not the major factor in the Müller-Lyer illusion (Fisher, 1970; Griggs, 1974; Massaro & Anderson, 1970). Thus, monocular viewing diminishes the visual Müller-Lyer illusion. But it does not abolish the illusion (Glennester & Rogers, 1993; Springbett, 1961). Furthermore, the verical–horizontal illusion also occurs with 3-D shapes (Heller et al., 2003).

Lateral inhibition, or the suppression of excitation in cortical neurons adjacent to neurons that are being activated, seemed a more promising explanation for the Müller-Lyer illusion. But the hypothesis also raised questions about effects of adjacent cues. Coren (1970) used Müller-Lyer configurations that excluded all adjacent cues. The usual connecting lines were omitted and the fins consisted of spaced sequential dots. The Müller-Lyer illusion was reduced, but it was still significant in these conditions. Lateral inhibition may thus also contribute. But it does not explain the Müller-Lyer illusion either (Gregory, 1998).

The heroic 1500 repeated tests with visual Müller-Lyer shapes to which Judd (1899, 1902) subjected himself convey some idea of the involuntary persistence of the visual illusion. But Judd also showed that the illusion was reduced, and finally disappeared, with repeated trials, suggesting practice effects. Periodical inspection of the data showed Judd that 750 repeated trials had reduced his 10-mm (18.5%) underestimation of a 54-mm Müller-Lyer figure to 2 mm (3.7%). The illusion increased briefly after test intervals for shapes that differed slightly in the length of the shaft and/or in the angle of fins relative to the shaft. But the further trials reduced the illusion much more quickly, suggesting that the reductions transferred across shapes. The total 1500 tests eliminated the illusion altogether, even producing a slight reversal.

A second observer, who knew nothing about the illusion or the aim of the study, produced comparable results. According to Judd, continued practice produces progressively better perceptual organisation, possibly by altering methods of inspection, although the perceiver is not aware of such changes during trials.

The justly famous study by Rudel and Teuber (1963) took the subject further by demonstrating that reductions in the Müller-Lyer illusion transfer across vision and touch. Twenty sighted adults were asked to centre the middle fin of adjacent shafts bounded, respectively, by divergent and convergent fins, so that the shaft was equal on either side. Ten participants were tested blindfold with tactual Müller-Lyer shapes for 80 consecutive trials. They then made six settings of the shapes in vision. The other ten participants were tested with the same shapes in vision for 80 consecutive trials. They were then blindfolded and made the further six settings of the shapes by touch. The design avoids the intermittent visual stimulus information that repeated cross-modal trials with the same participants involve. The finding – that the initial degree of the illusion was similar in the tactual and the visual condition – is thus all the more important. Moreover, the initial 80 consecutive trials produced significant, and roughly equal, decrements in both the visual and tactual illusions. Even more impressive, the decrements in each case transferred significantly from vision to touch and, somewhat more, from touch to vision in the subsequent trials in the other modality.

Reductions in the illusion with repetition across vision and touch could not be explained purely by low-level structural or retinal factors, such as hue, fundus pigmentation, or spatial frequency (Carrasco, Figueroa, & Willen, 1986; Ebert & Pollack, 1972). Indeed, low-level effects would not explain how or why practice reduces the Müller-Lyer illusion (Day, 1962; Judd, 1902, 1905; Malone, 1971; Rudel & Teuber, 1963) at all.

Higher-order perceptual learning and strategic-cognitive factors were invoked instead. The Gibson theory that perception depends on detecting higher-order amodal invariants (Gibson, 1962, 1966) suggests that perceptual learning produces progressively better differentiation of constituent features of shapes that are initially perceived as global configurations (Gibson, 1969). Tsai (1967) assumed that the tactual, as well as the visual, Müller-Lyer illusion is due to an initial global impression of the shape. As noted earlier, holistic impressions of new shapes are actually much less likely in active touch, if only because it depends on sequential scanning (Millar, 1997). Some effect of modality-specific aspects of inputs may be expected for unfamiliar shapes. It seems reasonable to assume that these have less effect and are less relevant to recognition once shapes have become familiar.

At the same time, the fact that repeated trials diminish an illusion is not itself evidence for a particular explanation of the reduction. Some reduction in the visual illusion also occurs in the first minute of inspection, without repeated trials (Predebon, 1998). In principle, Piaget's (Piaget & Inhelder, 1956) theory can explain that. The Müller-Lyer illusion is assumed to be due to shape features that elicit what he called undue "centrations", or involuntary fixations on or attention to given features. It is a primary illusion, according to him. It diminishes with age as children develop more systematic exploration and "centrations" become more balanced. Piaget only tested age effects in vision. However, the visual illusion certainly occurs quite strongly in adults. Over (1967) found no age effects for the haptic Müller-Lyer illusion. The notion of shifts in "centrations" is closer to the idea of shifts in attention from fins to shaft with repetition than to modality-specific explanations. However, Piaget (1953, 1954) assumed that perception depends on sensorimotor activities and that perceptual activities themselves are always tied to a sensorial field (Piaget, 1969).

Perceptual activities, such as eye movements in vision (Bolles, 1969) and finger or hand movement in touch, clearly differ. But neither explains the illusion in the respective modality. Finger and hand movements are essential in exploring tactile shapes. But different movements, as such, are not explanations. It is necessary to know the reason for using a specific movement rather than a different one, or of moving in a particular direction. Such factors – as well as the relation of a specific exploring or scanning movement to its start and goal location, and its possible spatial context – need to be specified beforehand and assessed independently. That has rarely been done.

The hypothesis that movement control mechanisms mediate the similarity of the illusion in vision and touch has also been disputed on the grounds that the degree of excursive eye movements had no effect on the haptic illusion, or vice versa (Wong, 1975). There is also evidence that the Müller-Lyer illusion occurs in passive touch, when the shape is pressed into the palm (Révész, 1934). The illusion in these conditions does not involve scanning movements at all. Similar findings show that eye movements do not explain the visual Müller-Lyer illusion either (Gregory, 1998). The visual illusion occurs as much, or more, when fixating a small Müller-Lyer shape without eye movements.

There is, however, one movement hypothesis that did need to be checked out specifically. It suggests that exposure or exploration time is the common factor in the visual and tactile illusion. Visual Müller-Lyer shapes with divergent fins needed longer exposure times than expected, whereas the exposure time for shapes with convergent fins was shorter than expected (Erlebacher & Sekuler, 1969). It has also been suggested that eye movement saccades take longer for divergent than for convergent figures (Festinger, White, & Allyn, 1968; Judd, 1905). At least one study has reported correlations between eye movements and pointing movements that were screened from sight (Moses & DeSisto, 1970). But timing data have not always been very precise.

It is known that movement time affects magnitude estimates in touch (Hollins & Goble, 1988) as well as in vision. The possibility that exposure or exploring times may underlie the illusion in the two modalities could thus not be discounted. But precise data on latencies for scanning movements in haptic convergent and divergent figures were needed to test the hypothesis that tactile Müller-Lyer shapes with divergent figures are overestimated because they take longer to inspect or scan, whereas shapes with convergent fins take less time to inspect or scan. The results of an experiment that tested the hypothesis are described below.

Another possible common factor is suggested by the hypothesis that the visual illusion depends on the distinctness of the constituent fin and shaft features. The "distinctive cue" hypothesis summarises a number of findings with the visual illusion. The studies produced changes in the size of the visual illusion by varying the length of shafts or fins, or by using distinctive colour cues for fins, or by adding internal or surrounding cues. Some presented the shaft and fins at different temporal intervals. Each differentiating cue reduced the illusion to some extent. But no single differentiating cue has been shown to eliminate the visual illusion completely.

Furthermore, telling people to ignore the fins was not effective on its own (Anii & Kudo, 1997). Additional discriminative cues seem to be needed (Coren & Porac, 1983). Explanations varied from confusion about the end points of the shafts, effects of contextual or conflicting orientation cues, to differences in attention to shaft and fins (Coren, Girgus, & Schiano, 1986; Coren & Porac, 1983; Dewar, 1967; Fellows, 1967; Fraisse, 1971; Girgus & Coren, 1973; Goryo, Robinson, & Wilson, 1984; Pressey, 1974; Pressey & Pressey, 1992).

The various interpretations are not necessarily mutually exclusive (Predebon, 1996; Wenderoth, 1992). The point is that the findings all depend on manipulating the discriminability of the fins from the shaft (Mack, Heuer, Villardi, & Chambers, 1985). That could well apply also in touch.

The "distinctive cue" hypothesis consequently also needed to be tested in touch. Other conditions that had not been previously been checked out in touch or vision were the effects of external and body-centred reference cues.

A series of experiments were therefore run (Millar & Al-Attar, 2002). Repetition effects were avoided by using only four trials in any one condition and allocating participants randomly or matching them for different conditions. The conditions used for testing, and their implications, are discussed separately for different hypotheses in the sections that follow.

MOVEMENT TIME AND DISTINCTIVE FEATURES IN THE HAPTIC ILLUSION

The question was whether we should explain the similarity of the illusion in touch to that in vision by the time it takes to explore convergent and divergent figures, or by the relative distinctness of fin-to-shaft features in touch as well as in vision, or possibly both.

Studies on vision had shown that a single differentiating cue diminished the illusion, but did not abolish it. Another differentiating cue that diminished without eliminating the illusion was the length of the fins. To make the possibility that these cues also affect touch doubly sure, we combined the two differentiating cues in the experimental stimulus set and compared the outcomes with the results of a control set of stimulus shapes.

Two different sets of horizontal and vertical raised-line Müller-Lyer shapes were prepared on transparent sheets and fastened to transparent plastic squares. In the control set, the length of fins was in the most commonly used proportion to the shaft length, and both shaft and fins consisted of plain raised lines in both divergent and convergent shapes (Figure 7.1). In the experimental set, only the shaft consisted of a plain raised line. The assumption that the illusion is due to the distinctness, or lack of distinctness, of the fins was tested additionally in all experiments. The fins in the experimental set consisted of smaller raised-dot fins that produced a textured feel that differentiated them from the plain line of the shaft, and they were also only half the size of the fins in the control set (Figure 7.1).

All test sheets in both sets of horizontal figures contained a divergent and a convergent figure. The two sets contained equal numbers of convergent and divergent shapes as either standard (top, scanned first) or as the comparison (bottom, scanned second) shapes. The locations of standard and comparison shapes in each set were counterbalanced across participants and runs in the relevant condition. The task was always to adjust the length of the shaft of the comparison figure to the length of the standard by stopping or continuing scanning along the trajectory of the comparison shaft.

The video-recording and timing device described in Chapter 4 was used with horizontal shapes. The device records participants' finger and hand movements over the shapes from below transparent surfaces. Latencies (1/100 s) for all finger movements are read off in frame-by-frame analyses. The question was whether differences in movement time could explain the haptic illusion. The latency graph for scanning horizontal shapes (Figure 7.2) shows that only the first of the four trials showed any significant difference in latencies between shapes. Shapes with convergent fins took longer than shapes with divergent fins, but only for figures with large plain fins. Latencies for comparison figures (not shown here) were also higher on the first trial. But they did not differ between convergent and divergent figures. Shapes with

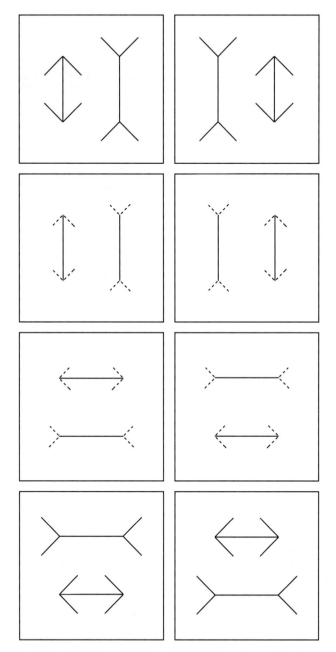

Figure 7.1 Vertically and horizontally oriented Müller-Lyer shapes with divergent and convergent fins. Shapes with plain, large fins were used as control sets. Shapes with small, textured fins were used as experimental sets.

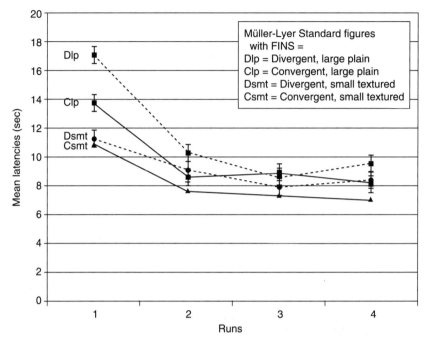

Figure 7.2 Mean latencies for four runs to scan horizontal Müller-Lyer shapes with divergent large plain fins (Dlp), divergent small textured fins (Dsmt), convergent large plain fins (Clp) and convergent small textured fins (Csmt). Divergent and convergent figures differ only on the first run. [From S. Millar & Z. Al-Attar, "The Müller-Lyer illusion in touch and vision: Implications for multisensory processes", *Perception & Psychophysics*, 2002, *64*, 353–363. Copyright © Psychonomic Society Inc.]

large plain fins took longer than shapes with small textured fins. But even that effect was found only on the first trial.

The analysis of mean constant errors (signed positive for overshoots and negative for undershoots) showed that the illusion was nevertheless highly significant for the experimental as well as for the control set of shapes. Movement time could not, therefore, be considered as a major element in producing the haptic Müller-Lyer illusion and could consequently not be regarded as a possible common factor in the tactual and visual illusion.

The experimental set of shapes with small textured fins had been designed to reduce fin effects by making the fins easier to discriminate from the shaft. That effect was significant. But the reduction in errors was only significant for shapes with convergent shapes. Underestimation errors were significantly larger for convergent figures with plain fins (−12%) than for convergent shapes with small, textured fins (−6.13%). But the difference in fin size and texture was not significant for shapes with divergent fins. Nevertheless, the haptic illusion that produced negative errors for convergent shapes was highly significant also for convergent shapes with small textured fins. The findings are graphed in Figure 7.3 together with the findings of subsequent experiments with touch.

The finding that distinctive fins reduced the illusion only for convergent figures, but had no effect on divergent figures in touch suggested a modality-specific difference between touch and vision. It could be explained by the scanning movements used for convergent shapes. Scanning the fins and shaft in shapes with fins that are angled towards the shaft brings the scanning finger into closer contact with the longer plain fins and plain shaft than is the case for the shapes with outgoing fins of the divergent set. The difference in fin effects between convergent and divergent shapes could, therefore, have been due to the proximity of fins to shaft in the convergent shapes. The proximity in scanning movements for convergent shapes would enhance the difference in feel between the shaft and distinctive fins in the experimental set with small, textured fins. The distinctiveness of fins from shaft thus had some effect also in touch, albeit only for convergent shapes.

However, the highly significant effect of the illusion, showing that convergent shapes were underestimated and divergent shapes were overestimated,

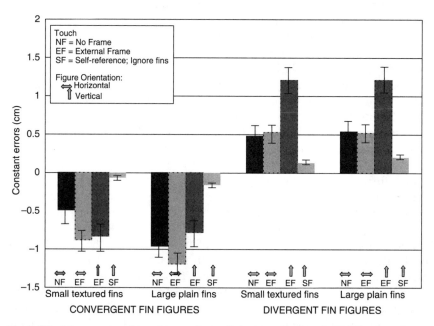

Figure 7.3 Mean constant (signed) errors for tactile horizontal and vertical Müller-Lyer shapes with convergent and divergent fins in NF (no frame), EF (external frame) and SF (self-referent) instruction conditions. [From S. Millar & Z. Al-Attar, "The Müller-Lyer illusion in touch and vision: Implications for multisensory processes", *Perception & Psychophysics*, 2002, *64*, 353–363. Copyright © Psychonomic Society Inc.]

showed that movement time could not be considered a major factor in reducing the touch illusion either.

Neither the movement time nor the distinctness of fin hypothesis could thus explain the haptic illusion or its similarity to the visual Müller-Lyer illusion.

SHAPE-BASED DISCREPANCIES IN LENGTH OR SIZE AND ADDED REFERENCE INFORMATION

It is proposed here that the very fin features that constitute Müller-Lyer shapes, and the overall size of the configuration, produce discrepant cues about the length of the shaft, which bias perception of the length of the shaft.

Tsai (1967) attributed the tactual illusion, as well as the visual illusion, to the global shape impression. That may be the case also in touch, even though haptic scanning with the fingertip cannot provide as immediate an impression of the outline of an unfamiliar shape as visual perception. Fellows (1973) showed that the "enclosing" nature of ingoing fins is important in the visual illusion for convergent shapes. It suggests that the same features that constitute Müller-Lyer shapes in vision also operate in touch. The discrepancies in length and size cues obtain in both.

The reference hypothesis assumes that, in principle, providing additional reliable reference information should override shape-based discrepant length cues and so reduce the illusion.

There is little previous evidence about the possible role of external reference information on the Müller-Lyer illusion in vision, although the effects of the spatial context on some other visual illusions has been demonstrated (Spivey & Bridgeman, 1993). However, the Müller-Lyer illusion clearly occurs in vision. It must therefore occur either despite the presence of external reference cues or, alternatively, because of the presence of environmental cues. External-frame information may even contribute to the visual illusion (e.g. Predebon, 1992), by increasing the effects of the apparent overall size of the configuration. In haptic conditions external cues are typically lacking.

The map study mentioned previously (Millar & Al-Attar, 2001) also included a Müller-Lyer configuration of a route within the map. Such a configuration does produce the illusion in visual maps (Gillan, Schmidt, & Hanowski, 1999). It also produced a significant Müller-Lyer illusion in the tactual map. Moreover, the combination of instructions to use body-centred reference cues and external reference cues from scanning the tangible surrounding frame with the other hand eliminated that Müller-Lyer illusion as completely as the inverted T-illusion (see Chapter 4).

In the case of the inverted T-junction, the complete elimination of the illusion had to be attributed at least in part to the map context in which it was tested. The dual instructions alone had reduced, but not abolished, the shape illusion without that context. A possible explanation is that instructions to use external reference in addition to body-centred reference has little or even adverse effects on shape-based illusions, unless the context makes them relevant.

It was, therefore, of interest to test the effects of external-frame instruction on haptic Müller-Lyer shapes when the task did not involve a map context. In a further experiment, naïve right-handed participants were therefore instructed to use their other (left) hand to gain information from a tangible external surrounding frame. No self-referent instructions were given this time. Participants were told to use information from the frame in judging the length of the standard shape and adjusting the shaft of the comparison shape to the standard. The procedures and conditions were otherwise the same as before.

The results for horizontal shapes with and without the use of the external frame are shown in Figure 7.3. The external-frame information produced no significant main effect. However, it increased the illusion somewhat for convergent shapes with small textured fins. It seemed possible that the mainly downward (vertical) movements for the frame relative to scanning the horizontally oriented figures could explain why the external frame had so little effect.

Two further experiments consequently used vertically oriented (sagittal plane) Müller-Lyer shapes (Figure 7.1), so that the main scanning direction of the shapes was aligned with the main direction of scanning the frame in tabletop space (Millar & Al-Attar, 2002).

Figure 7.3 shows that the Müller-Lyer illusions were also highly significant for the vertical convergent and divergent shapes. Moreover, with external frame (EF) instructions, both the convergent and divergent illusions were significantly larger than in the other touch conditions, suggesting that external-frame information is deleterious. The finding that using external reference cues does not reduce the Müller-Lyer illusion in touch, or makes it worse, is consistent with the fact that the illusion occurs in vision even though external background cues are normally present in vision.

The next question was, therefore, about effects of body-centred spatial reference. The hypothesis that accurate shape perception depends on congruent spatial cues from diverse sources suggests that additional spatial reference may override shape-based discrepancies. External-frame information seemed to increase the illusion, presumably by making the discrepancies in cues to the overall size more salient. Body-centred spatial reference could be more relevant to overriding discrepant length cues in the shapes.

In principle, of course, body-centred cues were present in all conditions. But the illusion occurred nevertheless. In fact, however, people are rarely aware of the use of body-centred cues for reference without experience and/or instruction (Millar, 1997). The same shapes were thus used, but the surrounding frame was not touched. Participants were explicitly instructed in how to use body-centred reference information in scanning the shapes. In addition, they were told to ignore the fins in using body-centred reference information. Instructions to ignore the fins evidently had no effect in vision on their own (Anii & Kudo, 1997). But they reduced the illusion to some extent when combined with other manipulations, although without eliminating the illusion (Coren & Porac, 1983; Goryo, Robinson, & Wilson, 1984). Instructions to ignore the fins were, therefore, added here to the instructions to use body-centred reference information. But the whole of each shape had to be scanned nevertheless also in that condition. The findings for all the haptic conditions are shown in Figure 7.3.

The results were rather startling. The haptic illusion was virtually eliminated with the instruction to use body-centred reference cues. For divergent shapes it was reduced from 15.13% in the external-frame condition to 2.06% with self-referent instructions. The reduction in underestimation for convergent shapes was from −10.06% with external frames to −1.44% with self-referent instructions. The differences are graphed in Figure 7.3, together with all results for touch with horizontally oriented and vertically oriented Müller-Lyer shapes with and without external frames compared to body-centred instruction conditions.

The illusion was, consequently, tested with the same vertical shapes in vision. The transparent test squares were placed on a bright coloured background so that the shapes (darkened from the back) and the surrounding square frames were clearly visible. External-frame information was therefore available in visual control conditions, as would normally be the case in visual conditions. In the control condition participants were told to ignore the fins to ascertain whether that alone would be sufficient to eliminate the visual illusion. Participants in conditions with body-centred reference information were also told to ignore the fins. But they were also explicitly instructed to use body-centred cues. They thus received the same self-referent instructions as participants in the comparable experiment on touch.

In the control (no reference instruction) condition, the visual illusion was highly significant for both convergent and divergent shapes, despite instructions to ignore the fins. The illusion was larger for shapes with the larger, plain fins than for shapes with small, textured fins. The difference was consistent with previous evidence that distinctive fins reduce, but do not eliminate, the visual illusion.

By contrast, self-referent instructions reduced the illusion very significantly, and indeed to floor levels, so that there were no fin effects (Figure 7.4).

More important, the instructions to use body-centred reference cues reduced the visual illusion to the same level as in touch, virtually eliminating the illusion in both modalities. There were small residual errors for convergent shapes (−0.12 & −0.16 cm, respectively) and for divergent shapes (0.17 & 0.15 cm, respectively). They differed from zero on one-sample t tests, but not between vision and touch.

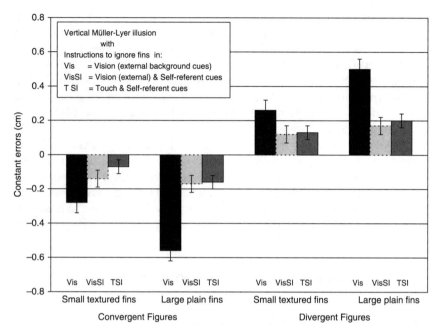

Figure 7.4 Mean constant (signed) errors for convergent and divergent Müller-Lyer shapes with small textured fins and large plain fins: in vision (external cues present) with instructions to ignore the fins (Vis); in vision with instructions to ignore the fins and to use self-referent (body-centred) cues (VisSl); and in touch with instructions to ignore the fins and to use self-referent (body-centred) cues (TSl). [From S. Millar & Z. Al-Attar, "The Müller-Lyer illusion in touch and vision: Implications for multisensory processes", *Perception & Psychophysics*, 2002, *64*, 353–363. Copyright © Psychonomic Society Inc.]

The almost complete elimination of the powerful Müller-Lyer illusion to the same level in vision and in touch, in the very few trials that were used here in all conditions, was new. Participants had been instructed that body-midline and body-posture cues could be related to the shapes. They seemed to have no difficulty in understanding the instructions.

We do not know precisely how individuals went about applying them. In principle, projecting midline and posture cues onto Müller-Lyer shapes in tabletop space provides reference anchors for the end locations of the shafts. The reference hypothesis implies that such location cues could override the biasing effects of discrepant size cues from the fins. The size of the reduction was unexpected. But the results were consistent with that analysis.

The Helmholtz principle that illusions are due to discrepancies in the very cues that normally produce accurate information applies. Müller-Lyer shapes are so constituted that the fin features give discrepant length signals relative to the shaft in two directions, and in relation to the size of the total configurations. The discrepancies in cues to length and size distort perception of the length of shaft in the direction of the overall size of the configurations. If anything, the bias is greater with external reference cues. External frames are more likely to relate to the total configuration, which may increase the bias.

The fact that body-centred instructions produced near-accurate size judgements in both touch and vision thus suggests that body-centred reference information is a common factor in shape perception by vision and touch that can override the discrepant length cues that characterise Müller-Lyer shapes in both touch and vision.

It does not follow that it is the only factor. The distinctiveness and length of fins, for instance, certainly affected the visual illusion. Distinctive small fins also had some effects in touch. But the salience of these effects evidently depended on the proximity of scanning movements of the constituents – that is to say, on modality-specific haptic conditions. Differences due to modality-specific conditions must, therefore, also be regarded as relevant to the illusion.

CHANCE OR SPATIAL PROCESSING OF SENSORY INPUTS?

The main motive for considering the three shape illusions in detail in the previous and present chapters was to see whether the similarity of illusions in touch and vision can be regarded as simply fortuitous. The evidence suggests that the similarity in the visual and tactual illusions is not a chance effect. All three illusions can be related lawfully to discrepancies in inputs with respect to different types of spatial reference. That is not to say that all types of illusion originate from exactly the same type of discrepancy.

The interpretation of the vertical–horizontal illusion, compared to the bisection illusion, is a case in point. Both the visual illusion and the tactual illusion could be explained by discrepancies in the relation of the vertical and horizontal length cues with respect to the vertical and horizontal extents of the delimiting visual field in the case of vision, and by discrepancies in the vertical (radial) movements versus horizontal (tangential) movements with respect to the total spatial extent of movements that are anchored, respectively, at the elbow and shoulder joints. The haptic L-illusion varies with the spatial plane in which it is presented. That needs to be checked out further. But the relative spatial distance rule that explains the visual illusion can also apply to the tactual illusion.

The bisection illusion, by contrast, is not orientation-specific. The discrepancy occurs within the shape. The point at which the continuous line meets the sectioned line acts as a boundary between the two smaller lengths, which make the continuous line seem longer in both touch and vision.

The startlingly similar effect of body-centred reference instructions on both the visual and tactual Müller-Lyer illusions is perhaps the best evidence we have so far for a common factor. According to Over (1968), a variable that affects haptic and visual illusions to the same extent can be regarded as common to both.

The findings that self-referent instructions reduced the Müller-Lyer illusion to near floor level in both touch and vision can therefore also be regarded as reasonable evidence that body-centred reference information is a major common factor in accurate shape processing by vision and touch, at least according to the Helmholtz principle that is espoused here.

Moreover, the finding that body-centred reference instructions alone reduced the tactual and visual Müller-Lyer illusions to near floor level, and that external cues had no or even deleterious effects, may explain also why the dual external and body-centred reference condition had less effect on the bisection illusion. External reference cues may have no effect, or may actually interfere with body-centred reference cues, when the bias is produced by discrepancies that arise from length cues within a configuration that affect perception of the overall size of the shape.

The two forms of reference were shown to be independent (Chapter 5). But how external cues affect other illusions is a separate question that requires further study.

The distinction between peripheral or "bottom-up" and central or "top-down" processes is clearly useful. But a dichotomy between "low-level" as opposed to "high-level" processes in explaining shape illusions is not justified. People are not generally aware of using reference cues or indeed of coding inputs spatially at all. If they were, explicit instructions to use reference cues could not have the very significant effects that they undoubtedly have. The findings here suggest that modality-specific effects do indeed contribute to the illusions discussed here.

It is by no means assumed here that all visual illusions occur in touch (Suzuki & Arashida, 1992; Wenderoth & Alais, 1990), nor that all haptic illusions occur in vision (Wenderoth & Wade, 1981). Even the three illusions discussed here differ clearly from each other. But they do suggest that different forms of spatial reference provide an important common factor for shape perception by vision and touch.

CHAPTER EIGHT

What does vision contribute to touch?

The discussions throughout this book foreground evidence on spatial coding in touch. There is no intention, and no danger, of losing sight (sic) of the fact that vision is the major spatial modality for humans, as it is for other primates. One can think of dozens of good biological reasons why that should be so. It is easy to cite the need to explore the environment for food and shelter and for cues that guide locomotion, anticipate obstacles, and, in the case of bipeds like us, help to achieve an upright posture. Nevertheless, as noted earlier, there is sufficient evidence to reject the extreme notion that vision is necessary for spatial performance, let alone that spatial processing can be equated with vision.

This chapter is concerned with recent suggestions that even "noninformative vision" improves haptic perception. The studies raise important questions, although the meaning of the phrase is not always clear. *Prima facie*, the term implies that vision affects touch even if it provides no relevant information. The reason for the extended discussion is that the studies raise the possibility that vision has wider and more pervasive effects on haptic processes than that of providing additional external spatial information. The possibility would have important implications for theoretical descriptions of the relation between touch and vision, and for rehabilitation in blindness and visual agnosias.

The notion of irrelevant spatial information is briefly considered first in the context of stimulus–response compatibility studies. The next section singles out two studies of noninformative vision precisely because they differ widely in tasks and experimental conditions. The details of what visual information was available and what was excluded in the studies are discussed to clarify the distinction that is being addressed. It is argued that the quite different visual cues that were available in the two studies could potentially provide relevant spatial information, albeit indirectly. Such processing differs from orienting to a common location. The distinction is valid. But it can be misleading if orienting to a location it is taken as the major criterion of what constitutes spatial processing.

A study that tested the possibility that vision may indeed have a more pervasive general effect on touch is then considered in detail. The discussion focuses on the implications of the results both for the relation between vision and touch and for theories of spatial integration.

STIMULUS–RESPONSE COMPATIBILITY AND IRRELEVANT SPATIAL CUES

The most obvious evidence that spatial aspects of vision can facilitate performance, even when the experimenter does not mention spatial cues and does not consider that spatial cues are relevant to the explicit purpose of a task, comes from "stimulus–response compatibility" studies. People react faster and more accurately when the stimulus that is to be discriminated from another stimulus appears on the same side, and in the same hemispace, as the relevant response key (Wallace, 1972). The response is faster and more accurate, whether colours, letters or numbers have to be discriminated. Studies of effects of this kind, including the "Simon" effect that occurs with auditory stimuli, have a long and distinguished history (e.g. Fitts & Seeger, 1953; Nelson & Greene, 1998; Nicoletti, Umiltà, & Ladavas, 1984; Proctor & Reeve, 1990; Simon & Rudell, 1967; Weeks & Proctor, 1990; Weeks, Proctor, & Beyak, 1995).

This is not the place to review the vast literature on choice reaction-time experiments that are primarily concerned with stimulus–response relations rather than with inputs from different sensory modalities. However, findings on task-irrelevant spatial cues are of interest in the present context.

The point is that tasks set by the experimenter, which seem to have nothing whatever to do with spatial information, are nevertheless performed better when the spatial location of the correct response key is on the same side as the location of the stimulus that is to be discriminated or is to be judged as the same or different from another stimulus. Performance is more difficult when response conditions are changed so that the "irrelevant" spatial location of the correct response key is inconsistent with the location of the critical stimulus. It is worse, for instance, if the correct response key has to be reached with crossed hands or is located in the contralateral hemispace to the stimulus location.

Moreover, the direction of response movements that is not intended to be relevant to the task – for instance, having to move the responding hand up to the right or down to the left – also produces orthogonal compatibility effects (e.g. Cho & Proctor, 2005), as do orthogonal stimulus and response positions (e.g. Ladavas, 1987). Lippa (1996) argued that the effect could not be explained by the usual assumption of a dimensional overlap so that the activated stimulus position automatically primes a similar positional response. On the other hand, accounts in terms of orthogonal coding, or preference for particular movements, could not explain the simpler effects. Lippa provided evidence instead that response positions are coded relative to the hand posture. The spatial reference account was shown to cover most stimulus–response compatibility findings.

It has to be concluded that, despite their apparent irrelevance to the stimulus–response task as set by the experimenter, the spatial location of stimuli, and the spatial location and direction of response activities that the experimental conditions require, are relevant to the performance of the task that confronts the participant, although the experimenter did not intend spatial position or direction to be a relevant factor, and even if the observers are quite unaware of being influenced by spatial cues.

WHEN AND WHY IS SPATIAL VISION "NONINFORMATIVE"?

Studies of "noninformative" vision are about the relation between inputs from diverse sources, rather than about stimulus–response effects. The question of why and how apparently irrelevant added vision improves haptic processing thus has to focus primarily on the details of the information that is available in the input conditions. Since inputs from vision and from active touch vary for different tasks, task conditions are necessarily important.

The main motivation for recent studies of noninformative vision has been the considerable resurgence of interest in crossmodal and multimodal perception (see Calvert, Spence, & Stein, 2004). The task conditions that have been used in the studies considered here are thus of particular interest also for that reason.

The first study to be considered provides a restricted meaning of "noninformative vision", namely vision that is irrelevant to the haptic task that someone is asked to perform (Newport, Rabb, & Jackson, 2002). The study specifically tested predictions from sensory integration models. The implications of such models are similar to the proposal that the form of spatial coding depends on the diverse inputs that are available for processing as reference cues in a given task (Massaro & Friedman, 1990; Millar, 1994), although there is some difference in emphasis. Optimal integration models of sensory integration (e.g. van Beers, Sittig, & Gon, 1999; van Beers, Wolpert,

& Haggard, 2002; Ernst & Banks, 2002) emphasise the salience and precision of sensory features that determine how the modalities are integrated. Computational models of optimal integration assign precise weights to the salience and precision of different sensory inputs. That makes it possible, in principle, to use approximate and non-linear mathematical functions to model interactions and to make testable predictions. Implicit but reasonable assumptions about the relative strength of visual and proprioceptive cues, assumed to convey information about spatial directions, predict the most efficient integration (e.g. van Beers, Sittig, & Gon, 1999). Computational models specify the relative strength of stimuli. But they do not always specify explicitly how the salience of the sensory inputs relates to the relevance or irrelevance of their contribution to detailed task conditions.

The experimental conditions in a study of noninformative vision by Newport, Rabb, and Jackson (2002) are of particular interest. The authors used a haptic orientation task. Participants used the left hand to feel the orientation of a standard bar that was located on the left of their body midline. They were to use the right hand to rotate a test bar, located in right hemispace, to parallel the orientation of the standard bar. The task was thus clearly spatial. In the control condition participants were blindfolded so that they had no current visual information.

To test for effects of noninformative vision the authors used an opaque screen at shoulder height, which prevented participants from seeing the test materials or their hand and arm movements in manipulating the oblique bars. But the participants had full vision of the screen and of the test area above and beyond it. Newport and colleagues imply that visual cues from these backgrounds are irrelevant to the task and are noninformative.

They tested the hypothesis that reducing the salience of proprioceptive cues would produce better performance with noninformative or irrelevant vision. The spatial locations of the standard and test bars were therefore at different distances from the body midline. In effect, the experimental conditions were heavily weighted against the usefulness of body-centred midline cues as reference cues for perceiving oblique directions accurately by touch and movement. The test bar that was to be rotated was always located at 10 cm horizontally to the right of the participant's mid-sagittal (midline in tabletop space) and 10 cm forward from the start position of the hand. By contrast, the standard bar was presented at four different locations, none of which matched the location of the test bar. The location of the standard bar was either nearer or further up from the start, and further out or nearer to the mid-sagittal line, than the test bar. At the four locations of the standard bar, four oblique orientations (angled at 20 degrees and 60 degrees clockwise and anticlockwise, respectively) were presented and had to be matched by rotating the test bar.

The authors showed that haptic performance was more accurate with the apparently irrelevant vision of the external background than in the blind condition that excluded all visual cues. The effect of the different length of movements or distance of the four bar locations from the mid-axis was reported separately. Errors, showing underestimations of the angle of orientations of standard bars, decreased consistently and proportionately from anticlockwise to the clockwise oblique bar orientations. The slope was identical for the vision and no-vision conditions. The authors, in my view rightly, dismiss the notion that systematic haptic errors warrant the notion that "haptic space" differs from "visual space". In terms of optimal sensory integration models, less reliable proprioceptive inputs affect the relative weighting of the two inputs. The authors thus attribute the advantage of the vision over the no-vision condition to the greater salience of extrinsic eye-centred or world-centred positional cues when proprioceptive cues for position information are rendered unreliable.

However, it is not clear why the authors identify vision solely with extrinsic, eye-centred or world-centred cues. Body-centred cues, as well as external cues, are present and important in vision, although we are not necessarily aware of using them. But we habitually relate environmental visual cues to vertical and horizontal directions, to each other, and also to body-centred posture cues, albeit not always very precisely. These include cues from the neck and head position, as well as gravitational cues to vertical and horizontal directions (Bottini et al., 2001; Snyder et al., 1998). Body-centred reference cues are in common between many visual and tactile conditions (Chapter 7). Proprioceptive cues from head, neck and body postures and gravitational and body-midline cues, as well as visual cues, are involved in sitting upright. The combination of external background cues with body-centred midline cues for vertical directions in vision would thus strengthen the effectiveness of haptic body-centred midline inputs as reference cues for hand and finger positions, even when there is no direct vision of hand or finger positions. Relating external vertical cues to common body-midline cues in vision and touch explains how visual information about external cues can make the use of body-midline reference cues more effective for coding orientations at discrepant locations. The effect of external spatial cues from vision would be indirect, but by no means irrelevant.

The point is that screens which prevent people from seeing their hands manipulating the orientation of a bar or a shape, but allow them full vision of the space above and beyond the screen, do not prevent the integration of external visual reference cues with body-centred reference cues. On the contrary, the combination of compatible external and body-centred cues enhances effects of reference information on touch, as shown by the findings discussed in Chapter 5.

Reference information is necessary for perceiving and adjusting the orientation of bars or shapes. The integration of reference cues is even more essential when task conditions are designed to make it more difficult to use the body-centred midline cues from touch alone. But orientation tasks cannot be performed accurately at all without reliable reference information.

The analysis of the (Newport, Rabb, & Jackson, 2002) task conditions thus suggests that the visual information from the external environment beyond the screen cannot be considered noninformative in the sense that it was irrelevant or could provide no useful information for the task. The use of a screen makes the visual cues less direct, because it requires the integration of visual cues with common gravitational and other body-centred reference cues. But, if anything, task conditions that make it more difficult to use body-centred cues alone make the use of external visual cues more relevant.

The findings support the sensory integration model that the authors set out to test. However, a shift in focus may be needed from predictions about the salience of sensory cues to foregrounding instead the spatial reference cues that the task undoubtedly requires, and which spatial processing of the diverse inputs provides, whether or not they are specified in the task as described by the experimenter to the participants. The authors actually show quite decisively in a control experiment in the same study that the perceived relevance of the visual, or of the haptic, information to the task determined whether the inputs had facilitating effects.

The task conditions in the other experiment (Kennett, Taylor-Clarke, & Haggard, 2001) are even more intriguing. The authors used a form of Weber's two-point threshold task to test tactile acuity. The threshold is the smallest distance between two points of a protractor touching the skin that are felt as two separate points rather than as a single touch point. Tactual acuity of the skin differs at different body sites and varies with the distribution of touch receptors. Acuity tests are not generally considered as spatial tasks at all, but simply as tests of sensory resolution.

It may thus come as a considerable surprise that tactile acuity judgements were found to be more accurate when people looked at the part of their arm that was going to receive touch stimuli. Even more striking, acuity was better still with magnified vision of the location of the arm (Kennett et al., 2001). The authors argue that the spatial information from seeing the patch of arm was irrelevant and noninformative, because vision alone, without the tactile stimulation, produced purely chance effects. They carefully ruled out the possibility that the visual enhancement was due to an automatic visual orienting response, by holding participants' gaze direction constant. They argued that the lower tactile threshold with vision could not be explained by increased spatial attention. Although the participants had been instructed to direct their attention always to the relevant arm location, a control experiment showed that looking at a (mirror-reflected) neutral object at the same location on the arm did not affect tactile acuity.

The authors attribute increased tactile acuity with vision to an online reorganisation of tactile receptive fields. It may involve bimodal cells that are sensitive to both vision and touch, or receptive fields that keep visual and tactile inputs about body-parts in close register (Graziano, Cooke, & Taylor, 2000; Graziano & Gross, 1993, 1994; Gross & Graziano, 1995; Rizzolatti, Fogassi, & Gallese, 1997). Feedback from brain areas that process inputs from multisensory sources may be involved. Evidence for temporary reorganisation of visual areas by tactile events (e.g. Maculoso, Frith, & Driver, 2000) is invoked to suggest that such reorganisation could underlie visual enhancement of tactile acuity.

At the same time, a purely on-line reorganisation of touch by vision would hardly explain the effect of artificially magnified vision. Why should magnified vision of the arm location make two separate, but closer, touch points feel further apart than with normal vision? Kennett, Taylor-Clarke, and Haggard (2001) mention the close relation of vision and touch with respect to the body and body-parts. They rightly stress the importance of attributing the effect of vision to seeing the skin of a body-part (Tipper, Lloyd, Shorland, Howard, & McGlone, 1998), and not to an irrelevant object at that site. However, the very fact that such attribution to a body-part is assumed to be the important factor in more accurate responses also implies that the information provided by seeing one's own arm was relevant to the judgement task, rather than irrelevant.

Craig and Johnson (2000) argued that the tactile two-point threshold is not actually a good test of tactile spatial resolution. Judgements at the same body location, and by the same person, vary too widely. Such variability suggests that results depend to a large extent on the criteria that participants adopt for what is to count as two points. Moreover, repeated trials make two-point judgements more accurate. Even more important, repeated trials lower the two-point threshold also on the other arm that has not been stimulated or tested at all. Such findings imply that learning or some form of "higher-order" processing is involved. Kennett and colleagues also assume that the effect is due to feedback from central processes. An assumption of feedback from central processes does not exclude criterion (β as well as acuity or d': Green & Swets, 1966) changes in processing the information. It does not exclude either that some form of spatial coding may be involved. Even the spatial direction of two-point stimuli on the arm can affect tactile threshold judgements (Richardson & Wuillemin, 1981). Spatial processing cannot be identified solely with coding extrinsic locations.

By considering the task conditions more closely, it is possible to guess at the kind of central processes that might be involved additionally, and what aspects of visual information might increase the accuracy of two-point judgements and might do so even more when the visual information is magnified. Two-point judgements do not require finding a spatial location. But they could involve spatial judgements. Two touch points on the skin are presented concurrently at varying but decreasing distances from each other, interspersed with a single touch stimulus. The person has to judge whether one or two points on a patch of his or her arm are being stimulated. Threshold responses fall at the smallest distance between the two points that the person can just still judge to be separate points. Below that distance, the person can no longer distinguish the two points from a single stimulus.

Strictly speaking, two-point judgements could, therefore, involve judging whether there is a distance between the felt touch cues. A pre-view of the patch on one's arm that is going to be stimulated is bound to be more relevant to judging whether there is a gap between two felt points than seeing a totally irrelevant object at the same site, or judging visual distances when there are no concurrent feels. Not being able to see the stimuli that touch a location on your body normally makes one look at the site. Preventing that orienting response may well elicit attempts to envisage it. Intelligent sighted adults who see the relevant patch of their arm just prior to repeated stimulation with points that vary in distance from each other, interspersed with single point trials, may well try to imagine whether there is distance between points that touch the skin. Visualising the distance between felt points to judge whether they are separate, especially if that area is magnified, could enhance haptic judgements of distinctness, as in other crossmodal conditions. The effect would be indirect, but not irrelevant.

Orienting to the source of stimulation from whatever source is indeed a basic response. Orienting visually to the source of a sound is an automatic response even in newborns (Wertheimer, 1961). Looking directly at the place on the body where touch is felt is more or less automatic also in adults. Such orienting responses are eminently useful biologically in maximising sensory inputs from a new or possibly threatening source (e.g. Cohen et al., 1997). The study thus addresses an important, as yet unsolved, problem about the mechanisms that could involve direct effects of central processes in changing tactile thresholds or acuity.

The suggested interpretations of the findings in both studies are, of course, pure conjectures. The point here is merely to suggest that participants could, in principle, have used the visual information that was available. If so, that could have conveyed task-relevant spatial information, albeit indirectly. In the first example, vision of the background beyond screens could provide extrinsic reference information for touch when mediated through body-centred reference cues that are common to both touch and vision. That would be highly relevant for accurate haptic manipulation of bar orientations. It is not, of course, necessary to suppose that the participants were aware of calibrating external and body-centred reference cues.

The conditions in the second example are consistent with other evidence that two-point judgements are not always simple tests of tactile spatial resolution. Knowing roughly what patch on one's arm is, or will be, touched does not exclude the possibility of envisaging, or imaging, the location in magnified form. That could convey associatively related, but relevant, information. Indeed, the authors of both studies suggest that the advantage of vision involved feedback from central or higher-order processes. That does not and need not exclude spatial processing of visual cues. There is considerable neurological evidence about the neural circuits that might be involved. The second study raises a different and as yet unresolved question about the mechanisms that may convey central effects to the periphery. The issue is briefly considered in the final chapter.

The main question is what is to be considered spatially relevant vision. Both studies took considerable care to prevent their participants from looking directly at the haptic manipulation and at the touch cue, respectively. The exclusion implies that only direct orientation of vision to the locus of manipulation or stimulation was considered to constitute spatial relevance. Visuo-spatial information that is integrated with haptic or touch stimuli via a common reference, or is mediated through a common context or representation, is indirect.

The distinction between direct and indirect connections between inputs from different modalities is valid. But it is not clear that connecting current visual inputs to current touch cues via central networks, or indeed via subcortical connections, makes the visual input irrelevant to the task or noninformative. It is debatable whether the distinction is best described in terms of informational relevance. To do so may be misleading.

The point in analysing these two quite different studies on the relevance of vision to active and passive touch is that they raise important questions about aspects of vision that actually reduce or exclude task-relevant information. The answer is not obvious. But it is needed if we are to understand how different aspects of vision affect haptic performance. It is also relevant to studies of the neural circuits that are involved in processing spatial and shape information, respectively, in dorsal and ventral neural streams.

The neurophysiological evidence shows that spatial actions which involve movement depend on multimodal processes in dorsal neural streams, whereas object recognition is associated with ventral neural streams (Andersen, 1999; James, Humphrey, Gati, Menon, & Goodale, 2002; Ungeleider & Mishkin, 1982). Studies with non-invasive techniques have also found that tactile tasks activate extrastriate, and possibly even primary, visual areas of the human brain. Effects vary with sensory experience and practice, consistent with performance in blindness (Büchel et al., 1998; Cohen et al., 1997; Sadato et al., 1996, 1998).

Moreover, activation by vision and touch in overlapping sub-regions in the ventral occipito-temporal cortex seems to depend on recognising geometric features of object shapes (Amedi, Jacobson, Hendler, Malach, & Zohary, 2002; Amedi et al., 2001; Pietrini et al., 2004). It suggests that such activation is associated with spatial aspects of object recognition.

The findings raise the intriguing question as to why these "visual" ventral areas respond to touch, and how spatial aspects of vision in ventral and dorsal streams relate to each other as well as to other aspects of vision that enhance touch and haptic performance.

A first step to answering such questions is to manipulate aspects of vision that actually provide relevant spatial information. That attempt was made in a study that looked at effects of different forms of impaired vision, which is considered in some detail in the next section.

WHAT ASPECTS OF VISION FACILITATE HAPTIC PERFORMANCE?

The possibility that vision has a general enhancing effect on touch even when it provides no relevant information is potentially important theoretically. It would also have practical implications not only for people who are born totally blind, but also for rehabilitating people with impaired vision and with visual agnosias. So far there is little evidence either for or against an assumption of such a pervasive effect of vision on touch.

To test the hypothesis that the mere presence of vision can enhance haptic performance required a spatial task precisely because spatial performance is usually better with vision. It also needed a visual condition that excludes indirect as well as direct visuo-spatial cues. The possibility that vision has quite pervasive effects implies that vision will produce greater accuracy in spatial tasks, such as remembering the spatial location of landmarks in a spatial map, even when the visual condition excludes all visuo-spatial cues.

The reference hypothesis, by contrast, makes almost completely opposite predictions. It implies that visual information benefits performance only to the extent that it provides cues that are relevant to a given task. For the task of remembering the spatial location of landmarks, it means that vision will benefit haptic performance only to the extent that it provides additional cues that contribute to the reference information that memory for spatial locations requires.

The combinations of touch with three typical forms of reduced vision were of particular interest in trying to understand what information vision actually contributes. Diffuse light perception presents a crucial test between the two hypotheses. In diffuse light perception, vision provides light but no shapes, edges, points, shadows, or even close finger movements, or visuo-spatial cues of any kind, within personal space. Diffuse light perception should improve accuracy in a haptic spatial task if vision has an enhancing influence on touch even when it excludes all visuo-spatial information. The reference hypothesis, on the other hand, predicts that diffuse light perception, which excludes all spatial cues, would have no advantage in a spatial task over touch without vision.

Peripheral vision and tunnel vision were two other well-known forms of reduced vision that were investigated. They were also of particular interest, because they reduce different aspects of vision. Peripheral vision completely excludes focal vision. Nothing is seen by looking straight down at the table, or at any object on it. Moving stimuli in the periphery can be detected, albeit not identified. Moreover, moving the head, neck and shoulders in various directions makes it possible to bring the somewhat blurred peripheral vision to bear on target stimuli and a swathe of surrounding object locations. In terms of spatial information, peripheral vision provides blurred vision of a relatively large space outside the central area. Visual features are not easy to discriminate clearly. They are, however, relatively easy to relate spatially to other cues and to parts of an external surround for reference.

Tunnel vision, by contrast, excludes peripheral vision, but looking directly at a shape provides clear, sharp focal vision, albeit of very small area with a diameter of 3.7 cm, as measured with spectacles that are specifically designed to simulate typical forms of residual vision. A small object is easy to see, and its location can be specified relative to body-centred cues. But the clear small patch of vision is not easily related to surrounding cues. By moving the head in different directions, the small clear visual area can be moved to other locations. But that also entails that the location of the original target is no longer in view. Thus, tunnel vision provides clear focal vision of a small central area. That makes it easy to discriminate small target features, but more difficult to use cues external to the target as reference cues to specify its location.

Given that both forms of vision add shape or spatial cues, some advantage for touch might be predicted, although presumably less for either than with concomitant full vision. The question was whether or not all forms of reduced vision have precisely the same effect on touch.

One method of examining that was suggested by the significant difference between target locations found in the study on additive effects of body-centred and external reference cues considered in a previous chapter. The finding was explained by unexpected distinctive touch cues near two locations, acting as further local reference cues. The interpretation needed to be checked out in any case with a new series of locations. It was thus a subsidiary hypothesis to be tested. But the method of using recall of a number of different locations was also expected to provide some indication of whether typical forms of reduced vision have identical effects on touch.

The map layout used in the studies contained six to-be-remembered locations along the designated route, marked by different shape symbols (Figure 8.1). Two sharp bends in the route, near Locations 3 and 5 respectively, were deliberately chosen to provide distinctive local touch cues. The point of doing so was to test the interpretation of the difference between locations that had been found previously (see Chapter 5). The subsidiary hypothesis to be tested was therefore that local touch cues act as additional reference points. It predicted greater accuracy for these two locations compared to the other locations. If so, that difference between locations should obtain in all conditions, including in conditions that combined touch with peripheral vision and with tunnel vision. The hypothesis would be disconfirmed if the six locations produced a serial position (bow-shaped) effect in all visual and touch conditions.

The main prediction from the hypothesis that vision has a pervasive vision effect on touch was that peripheral vision, tunnel vision and diffuse light perception would all improve recall accuracy, albeit to a lesser degree than full vision. Moreover, if peripheral and tunnel vision affect touch only in so far as both reduce visual information relative to full vision, they should not differ specifically in cue distinctiveness and reference information at different locations either.

The route and the shape symbols for the to-be-remembered locations in the map layout were embossed for maximum tangibility. They were also marked distinctly in black and white for easy visibility. The contrast of the light-coloured background of the map with the dark wooden tabletop also produced a visually distinct rectangular frame around the map.

Peripheral vision, tunnel vision, diffuse light perception and total blindness were simulated with professionally designed spectacles. Such spectacles are specifically designed so that mobility teachers and trainees can experience typical forms of reduced vision and total blindness at first hand. The use of specially designed spectacles here made it possible to allocate sufficient numbers of sighted people randomly to typical forms of diffuse light perception, peripheral vision, tunnel vision and total blindness to produce reliable effects. Effects on touch of different forms of residual vision had not been tested experimentally before. They were here compared with touch alone, using spectacles that simulated total blindness, and with effects of full vision on touch (Millar & Al-Attar, 2005).

Equal numbers of participants were allocated randomly to the five modality conditions. Touch was used in all conditions, either with full vision, or with the three forms of reduced vision, or without vision in the condition that simulated total blindness. Participants used their index finger to scan the six location symbols along the marked route. They were asked to remember the precise midpoint of the landmark symbol for each location along the route. The experimenter initially guided the participant's finger from the "start" to the "scan" point to show the scanning direction. The participant scanned independently from then on. The point was to place the midpoint of the ball of the fingertip on the midpoint of each landmark symbol as it was encountered in scanning the route, and to remember these locations. The experimenter named the next location in each case both in presentation and recall, so that neither the name nor the sequence of a landmark had to be

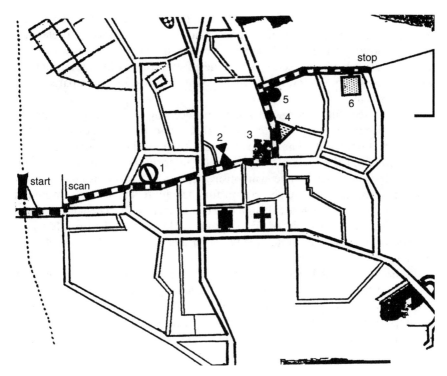

Figure 8.1 Tactile presentation map. Black areas are embossed; parallel lines symbolise roads. The designated route has guiding blocks. The symbols for landmark locations are numbered 1 to 6 here, though they were not on the actual map. The label "start" shows the start of guided scanning; "scan" shows the start of independent scanning. The test map was identical but without landmark symbols for locations. [From S. Millar & Z. Al-Attar, "What aspects of vision facilitate haptic processing?" *Brain and Cognition*, 2005, 59, 258–268. Reproduced with permission from the publisher. Copyright © 2005 Elsevier.]

remembered, only its precise location. In recall, the participant scanned the route from the "scan" point as before on a test map without the landmark symbols. The participant indicated each to-be-remembered location by placing the midpoint of the ball of the fingertip on the remembered midpoint of the landmark at that location. The error data were distances of that point from the actual midpoint of the shape at that location on the route, recoded on a map replica with gridlines, for every trial at each location.

The results from the overall analysis showed a highly significant effect of the four bimodal and the unimodal touch conditions. There was also a significant overall effect of the locations, as found in the study with a previous map (Millar & Al-Attar, 2004). But in this new map the location effect was due to greater accuracy of recall for the two locations that had been placed deliberately close to sharp turns in the designated route than for the other locations. The location effect supported the subsidiary hypothesis that distinctive local touch cues near target locations act as additional reference cues. The graph in Figure 8.2, which shows the comparison of touch with full vision and touch alone, also illustrates the effect of the distinctive local cue at Location 5, which was lower, but also present, at Location 3 compared to all other locations.

As expected, full vision with touch produced by far the largest improvement in accuracy compared to touch alone than did any other modality condition (Figure 8.2). The distinctiveness of the local cues from the shapes that symbolised the location was evidently important. The sharp turn that abutted the solid small circle symbolising Location 5 was clearly a very distinctive reference cue. The sharp turn for the square shape that symbolised Location 3 was a much less distinctive cue (Figure 8.2). The square shape had inadvertently been placed so that it fitted closely into the turn. Even so, Location 3 was more accurate than the other locations, except for Location 5, in all modality conditions.

It should be noted that accuracy for Location 5 did not differ between vision and touch, despite the huge advantage of vision otherwise. That was also the case in all other modality conditions. It incidentally also underlines the importance of distinct touch cues for haptic spatial coding. The role of distinct features is discussed further in the final section of this chapter.

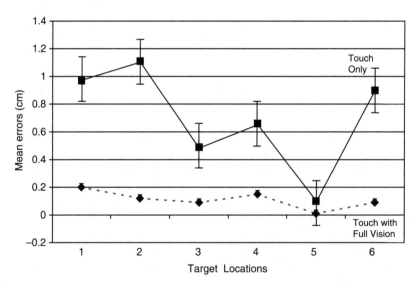

Figure 8.2 Mean errors for touch with full vision compared to touch alone. Note the lack of difference at Location 5. [From S. Millar & Z. Al-Attar, "What aspects of vision facilitate haptic processing?" *Brain and Cognition*, 2005, 59, 258–268. Reproduced with permission from the publisher. Copyright © 2005 Elsevier.]

Peripheral and tunnel vision added to touch were significantly less accurate than full vision. They were, however, also significantly more accurate than either diffuse light perception or touch alone. In fact, peripheral and tunnel vision produced the same level of accuracy overall. Both improved touch, although to a lesser degree than full vision. But their location effects differed significantly. That interaction was not due to Locations 3 and 5. These two locations showed the same relative accuracy in both peripheral and tunnel vision as in the other modality conditions. The significant interaction was due to Locations 2 and 4. The difference in effects on touch by peripheral and tunnel vision, and how both differ from touch without vision, is graphed in Figure 8.3.

The significant interaction of peripheral vision and tunnel vision with locations was consistent with observed differences as regards the clarity of feature discrimination and external reference cues, respectively. In tunnel vision the Location 4 was less accurate than in peripheral vision. The small triangle that symbolised the location would be seen as distinct from the adjacent angled line in tunnel vision. But the angled line would not provide a useful local reference cue, nor would body-centred reference provide accurate reference cues for the small shape. At the same time, head movements that transferred the clear area of focal vision to any part of the surround could, in

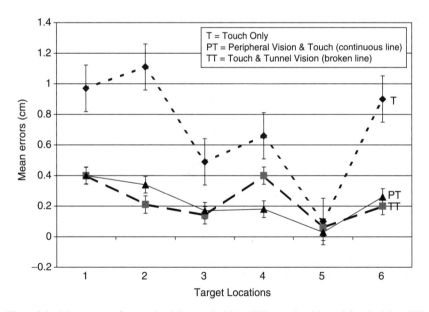

Figure 8.3 Mean errors for touch with tunnel vision (TT), touch with peripheral vision (PT) and for touch alone (T). [From S. Millar & Z. Al-Attar, "What aspects of vision facilitate haptic processing?" *Brain and Cognition*, 2005, *59*, 258–268. Reproduced with permission from the publisher. Copyright © 2005 Elsevier.]

principle, provide useful external references. However, such movements would have the effect of losing sight of the target shape and its location.

In peripheral vision the target shape and adjacent cue would be blurred. But the head movements bring the shape into peripheral view that could include the external (e.g. right vertical) edge of the map layout. That could, in principle, provide an external reference cue for the blurred target. The reverse location effect obtained at Location 2. The clear focal view in tunnel vision of the hour-glass shape at Location 2 relative to the horizontal line of the route would provide a local directional cue that would be more possible to relate to body-midline reference cues for recall. Blurred peripheral vision of the target shape and the abutting horizontal line of the route would make the shape less accurate as a location cue. Reference to concomitant external cues from the larger area or a vertical side should have some effect on accuracy, without necessarily compensating to the same relative extent for the tunnel-vision effect as at Location 2.

More empirical evidence on the precise effect of head movements is needed to determine whether the above interpretation describes the difference in location effects correctly. The experimental conditions had left participants completely free to use whatever head movements they needed to view the layout and target locations in all conditions, and they were observed to do that. What the findings clearly show is that although peripheral and tunnel vision produced the same overall amount of improvement relative to touch, they differed in opposite directions at two locations.

The interaction of peripheral and tunnel vision with different locations was consistent with an interpretation that attributed the effect, respectively, to the clearer feature discrimination in tunnel vision and an increased availability of external reference cues in peripheral vision. It implies that cue distinctiveness, as well as the integration of inputs with reference cues, has to be taken into account in characterising accurate spatial coding.

The crucial finding that distinguished the two main hypotheses hinged on the effect on touch of diffuse light perception, which provides vision but totally excludes spatial cues. The result supported the hypothesis that vision improves haptic performance only in so far as it provides relevant cues. Diffuse light perception was the only form of reduced vision that failed to show any advantage over touch, either in overall accuracy or at any location, including Locations 5 and 3 (Figure 8.4).

The location task required spatial cues, and that demand was the same in all modality conditions. But of the three types of reduced vision, only diffuse light perception excluded all spatial cues. It was also the only form of reduced vision that failed to improve haptic perception.

The results are not compatible with the assumption that vision improves touch even if it provides no task-relevant information. They imply that it is not the mere presence of vision which improves touch: The improvement

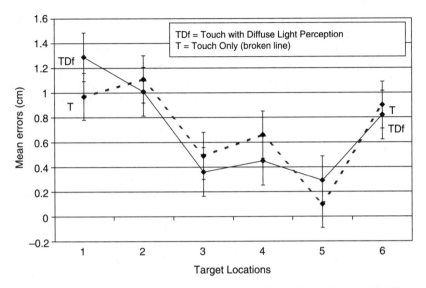

Figure 8.4 Mean errors for touch with diffuse vision (TDf) and for touch alone (T). The two conditions do not differ significantly at any location. [From S. Millar & Z. Al-Attar, "What aspects of vision facilitate haptic processing?" *Brain and Cognition*, 2005, 59, 258–268. Reproduced with permission from the publisher. Copyright © 2005 Elsevier.]

is due to the presence of cues that add to the processing of inputs for spatial reference that spatial tasks require.

It should be noted that all participants were sighted people with normal vision who were tested blindfolded. The fact that their haptic performance benefited from spectacles that provided peripheral or tunnel vision, but derived no advantage from spectacles that provided diffuse light perception, shows that the effect of added vision was due to current visual cues, not to previous visual experience or means of representation.

The findings on the bimodal and the unimodal tactual conditions were consistent with optimal integration theories. But they also suggest further that spatial and non-spatial aspects of vision need to be distinguished explicitly in bimodal studies and in theories of spatial integration.

Moreover, the results show that specific forms of residual vision combine differently with information from touch and movement. The interaction of location effects with peripheral and tunnel vision was consistent with the very differences between acuity of detection on the one hand, and the presence of potential reference cues on the other, that mark these two forms of reduced vision.

The role of sensory acuity in spatial processing is implied in describing spatial processing as integrating and organising diverse sensory inputs as the reference cues that spatial tasks demand. The location effects here suggest further that discriminative acuity and the use of local touch cues for reference also need explicit mention in predicting levels of accuracy.

The significant location effect supported the subsidiary hypothesis that distinctive local touch cues near target locations act as additional reference cues. It showed that distinctive cues are indeed used spontaneously for spatial coding. Perhaps the best indication of the importance of precision in detection and discrimination was the highly significant result for the location that was marked by a circle adjacent to, but distinct from, a sharp angle in the designated route. It was the most accurate location in all conditions, including in touch alone. More important, however, was the finding that the pattern of location errors for the new map did not replicate the bow-shaped curves found previously. The final location did not show either the "recency" or the "primacy" effects that characterise serial recall.

It is not assumed that the lack of serial effects here was due to the difference in modality between touch and vision or touch and hearing. Serial effects may well be found also with haptic information. They are likely to occur in experimental conditions in which people have to remember the order of sequential locations, That was not demanded here.

SUMMARY AND CONCLUSIONS

What does vision contribute that benefits touch? The assumption that even noninformative vision improves tactual performance raised important questions about the aspects of vision that do interact with and/or benefit touch. The assumption seems to imply that the mere presence of vision has a general ameliorating effect on tactual performance, even if it excludes all task-relevant information. The study that was designed to test the assumption directly did not support that hypothesis.

The detailed analyses of the experimental conditions in the two very different studies on "noninformative" vision suggested that the term may be misleading. The critical experimental condition in both studies was to prevent direct visual orienting, respectively, to the locus of the tactile stimulation and to the haptic activity of rotating a felt bar. But visuo-spatial information that could, in principle, be invoked in a two-point judgement task, and current visuo-spatial cues that could be integrated with haptic inputs in the manipulation task, were available, albeit via more circuitous central routes. There is an undoubted empirical distinction between orienting to a locus of stimulation and using available visuo-spatial information indirectly. But the distinction between orienting to a location and using spatial information indirectly cannot be used as a general criterion for what constitutes spatial information or for its relevance to a given task.

The wider question of whether vision has pervasive effects on touch even when it is irrelevant to the task was tested by comparing touch with full vision, touch without vision, and touch combined with three different types of residual vision in a task that required memory for the location of "landmarks" in a spatial layout.

The results showed that vision which provided no spatial cues was the only form of reduced vision that failed to improve touch at any location in the spatial task. Unlike peripheral vision and tunnel vision, diffuse light perception that excluded all task-relevant cues did not benefit touch. The findings imply that it is necessary to distinguish explicitly between spatial and non-spatial aspects of vision in considering relations between vision and touch.

The difference is not the same as the distinction between ventral neural stream object recognition and spatial actions that are associated with dorsal neural streams (Ungeleider & Mishkin, 1982). Deciding "what" an object is, or how it functions, can depend on its shape. Deciding whether an object is square- or diamond-shaped is an instance. There is evidence that geometric shape features of objects activate ventral circuits in tasks of object recognition (Amedi et al., 2001, 2002; Pietrini et al., 2004). The evidence implies that spatial processing of geometric features is involved in the activation of extrastriate visual regions. These include areas that are activated also by touch. But it is also possible that "object-centred" shape or spatial effects involve connections between ventral and dorsal processing streams.

It should also be said that the distinction between spatial and non-spatial vision does not mean that non-spatial vision has no biological functions or that it fails to combine with any other cues. The functions of non-spatial vision seem to relate primarily to conditions of arousal, mood and emotional states. Light perception is important for circadian rhythms, sleep mechanisms and hormonal effects in regulating the biological "clock" (Brainard, Rollage, & Hanifin, 1997). The primary functions of these mechanisms relate only very indirectly, if at all, to information-processing functions of inputs.

The evidence discussed in this chapter takes us some steps further into specifying what aspects of vision improve touch. The finding that vision did not improve touch when it excluded all spatial cues, in contrast to other forms of reduced vision, implies a functional difference that presumably relates also to differences in neural connecting circuits (Atkinson et al., 2001, 2003). The distinction is important. It could provide a relatively reliable baseline for studies of bimodal effects and explanations.

The fact that different types of reduced vision had different effects on touch was actually surprising. It would not have been expected from known effects of full vision. The findings were consistent with the criteria of spatial processing used here and with observed differences in the visual information that was available when wearing the spectacles that typified these forms of vision.

The results on peripheral and tunnel vision suggest more specifically that even within spatial vision at least two different aspects of information must be distinguished. Moreover, these aspects evidently do not combine in precisely the same respects with information from touch.

The results for peripheral and tunnel vision are potentially important for optimal theories of integration. Specific forms of residual vision evidently combine differently with information from touch and movement. It suggests that another factor needs to be taken into account in explaining the integration of information from vision and touch. However, further empirical investigations are also needed for practical reasons. Combining information from touch and movement with impaired or failing vision could potentially be used to enhance shape and spatial perception. But we need far more evidence on how these and other specific forms of residual or failing vision combine with information from touch.

The highly significant location effect highlights the role of distinctive cues in spatial processing. A role for sensory inputs is necessarily implied in describing spatial processing as the integration and organisation of diverse sensory inputs for reference. The integrating function of spatial processing has been foregrounded here as the main characteristic. Acuity in discriminating stimuli and the use of apparently irrelevant local touch cues for reference should probably be emphasised also in specifying the aspects of vision that combine with information from touch and movement and so add to inputs from touch alone in spatial tasks.

CHAPTER NINE

How far have we got? Where are we going?

Occam's Razor, the principle of parsimony, should perhaps be taken to apply only to the ultimate rules that may be found to govern biological systems. Two thousand years of progress and increasingly accelerating rates of empirical findings have uncovered ever more complex factors and interrelations between contributing factors rather than ever simpler accounts of human spatial behaviour. It seems nicely paradoxical that the aim of current cognitive neuroscience – to take account of findings from molecular biology (Zhou & Black, 2000), neurology, psychology and philosophical analyses – is only made possible by progressively more detailed studies of multiple small areas within the disciplines. But it does suggest the direction that is likely to take us forward.

I have focused on active touch, and what it may have in common with vision, in an attempt to clarify the role of sensory inputs in spatial coding. In this chapter, I shall try to assess how far the findings that have been considered are consistent with the predictions and also extend the assumptions from earlier studies, and what they may imply for future directions.

For convenience, the chapter is divided into sections that summarise the main results, their implications for the questions that were raised in the second chapter, and the further directions they indicate.

MOVING THROUGH LARGE-SCALE SPACE: INFERENCES FROM VEERING

The study on veering solved the puzzle of why people do not simply wander randomly in unfamiliar large-scale spaces that totally lack orienting cues, but veer from the straight-ahead more and more in a particular direction. The findings here showed that the heading direction is biased by unexpected and unrelated sound and posture cues in such spaces – that is to say, by stimuli from the very sources that help to maintain the direction of locomotion in spaces in which external reference can be integrated from the start with body-centred heading cues for reference and/or updating.

The surprising finding that the depth of alternate strides can also have biasing effects is further evidence for the effects of body-centred cues. But hearing an unexpected sound at unpredictable locations was just as misleading. The findings show that the absence of external reference cues that would normally interact with existing "heading" information rendered such heading cues liable to bias from the irrelevant sound. The sound was misleading, even though, or more probably because, sounds normally contribute to maintaining and updating the direction of locomotion.

The findings are consistent with explanations that stress the importance of the convergence or integration of stimuli from external and body-centred sources in providing accurate reference and updating cues. They also highlight the relevance of current perceptual cues in spatial coding.

The steep versus shallow sideways curves that alternate left and right steps in locomotion produce seemed to correlate with the speed of walking. The possibility that speed of walking may reduce postural or movement biases raises questions about the role of movement speed in relation to posture cues and to memory for kinaesthetic/postural cues that need further investigation.

FINGER MOVEMENTS AND SPATIAL CODING IN SMALL-SCALE SPACE

The review of findings on braille patterns showed that the form and functions of finger and hand movements not only get better, but actually change with experience, so that quite different tactual features are picked up in response to the demands of different reading tasks. The interrelations between factors that are likely to be involved are summarised in flowchart form in Figure 9.1.

The difference in the pick-up of features from lateral shear patterns in reading for meaning, compared to detecting a letter pattern, by early blind fluent braillists also raises important questions about the neural changes that this may involve. The neurophysiological evidence suggests that tactile braille and braille-like stimuli activate areas in occipital (visual) regions of the

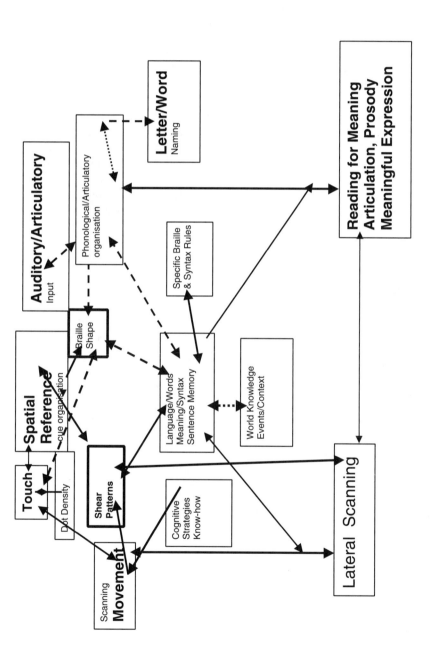

Figure 9.1 Flowchart showing connections between factors in braille letter-by-letter reading and lateral scanning in reading braille texts for meaning. [From S. Millar (2004), "Literacy through braille and perception by touch and movement", in Y. Eriksson & Kenneth Holmquist (Eds.), *Language and Visualisation*. Sweden, Lund University, published with permission.]

cortex more in early blind fluent braille readers than in late blind or sighted people.

The intriguing point is that the visual areas in the ventral stream that are colonised by touch in early blindness are normally associated specifically with visual object and shape perception. The pick-up of braille features differs particularly with long-term braille experience by the early blind. The question is, therefore, whether crossmodal compensation in blindness depends specifically on the pick-up of shape information from touch, or on the similarity of the perceptual features in vision and touch, or on how the constituent features relate to each other in a given stimulus array.

Moreover, an area of the extrastriate visual cortex has been shown to be active during tactile judgements of orientation also in the sighted (Pascual-Leone, Theoret, Merabet, Kauffmann, & Schlaug, 2006; Zangaladze, Epstein, Grafton, & Sathian, 1999). The cells in occipital areas that are activated by touch are specific to vision and touch, and do not respond to auditory signals in sighted adults (Amedi et al., 2002). A question that needs further research is whether the "colonisation" by touch of visual areas depends solely on the proximity of tactually sensitive neurons to the visually sensitive neurons that normally constitute the majority of neurons in these occipital regions of the cortex.

A recent study (Lewald, 2007) found that, in a head-pointing task, quite brief periods (90 min) of light deprivation produce more accurate sound localisation, which was eliminated after 180 min of re-exposure to light. The deprivation only reduced constant (directional) errors. The result was attributed to the absence of visual calibration in neural representation during light deprivation, rather than to a general compensatory reorganisation process. It is of interest in that connection also that far greater plasticity than was previously supposed has also been found in the adult human brain (e.g. Hummel & Cohen, 2005; Kujala et al., 1997; Steven & Blakemore, 2004).

We have come a long way since the discovery that initially endogenous brain activities are modified during development, with enriched versus impoverished early experience. Such changes could reasonably be assumed to correlate with improved performance due to greater stimulation and exposure to the more, and more diverse, information that is afforded by enriched environments. But more startling, at least to me, was the finding that the whiskers of cats and mice grow longer and thicker in animals that have been deprived of vision from birth (Rauschecker et al., 1992).

The temptation is to regard such findings as evidence for purely automatic compensation for the loss of a sense modality. Rauschecker and his colleagues were careful. They noticed an increase in the animals' normal use of their whiskers for spatial orientation, and they found a concomitant enlargement of the somatosensory cortical barrel field in binocularly enucleated mice. The authors proposed that increased use of whiskers for spatial orientation by visually deprived animals could stimulate growth beyond the normal and could expand the somatosensory cortical barrel field via activation of the respective neural pathways.

Crossmodal take-over with decreased stimulation of one cerebral region and increased stimulation of another is not confined to touch and vision. Superior auditory spatial tuning in blind compared to sighted people, and improved sound localisation that varied with the degree and extent of blindness, has also been shown (Lessard, Paré, Lepore, & Lassonde, 1998; Röder et al., 1999). Some findings suggest that neural circuits which connect to other areas of the cortex are involved. A study that used event-related potentials showed functional reorganisation of auditory attention by congenitally blind people in dichotic listening tasks. Interestingly, the findings on the blind suggested a progressive recruitment of parietal (spatially functioning?) and then occipital regions (Liotti, Ryder, & Woldorff, 1998). Bimodal neurons in the monkey's anterior parietal cortex evidently retain information from visual or auditory cues that had previously been associated with haptic information. They seem to be activated as, and when, the information was needed for performance of the task (Zhou & Fuster, 2004).

It is not clear as yet how far the take-over of visual or auditory receptive or associative fields in the total, or transient, absence of a modality-specific input can be attributed to automatic sensory compensation. The compensation is clearly not complete, since young blind children need a good deal of experience and training to make full use of auditory and tactual information. Sathian (2000) was right to suggest that, at least with braille, practice makes perfect. The benefit of relevant experience (Goldreich & Kanics, 2003) is likely to be facilitated if lack of vision also produces crossmodal compensation in other modalities automatically.

An important direction for future work is to devise behavioural tests that isolate the specific factors that may differ in apparently similar tasks. It would serve to distinguish these specific effects from factors that have a general compensatory, or else deleterious, effect. Such detailed experimental analyses are essential if we are to unravel the factors that underlie some of the questions that the behavioural findings have pointed up for further neurophysiological investigations.

A related problem is whether, or to what extent, practice and experience could reverse or bypass the changes in behavioural and neural (synaptic) connections that are due to lack of early sensory stimulation. What criteria establish that the changes are totally irreversible, or cannot be by-passed, if sufficient subsequent stimulation or supplementary experience is provided over time?

If anything, the question of automatic compensation arises even more for touch from findings that suggest heightened tactile acuity of the braille reading finger (e.g. van Boven et al., 2000). There is probably enough evidence to suggest that the expansion of the representation of the reading finger in the somatosensory cortex comes about with increased stimulation (Pascual-Leone & Torres, 1993). But a further question for the future is how changes or expansions of cortical and/or subcortical neural representations and/or neural circuits affect tactile acuity thresholds. Do they depend on criterion changes? By what mechanisms can the expansion of cortical neurons and/or changes in synaptic connections have direct effects on the receptor organs in the finger? There are a number of further questions also about the molecular changes at the level of skin receptors with increased or decreased stimulation, and whether or how these relate to changes at cortical and subcortical levels.

These are issues for the future. More immediate questions for me concern the behavioural methods that can distinguish between specific "top-down" effects. For instance, we know relatively little as yet about the factors that produce changes in finger movements in the pick-up of haptic perceptual features for different braille reading tasks. In what respects, for instance, do they resemble changes in eye movements for different visual reading tasks?

The role of scanning movements in coding of braille and larger layouts spatially was considered in relation to the function of the two hands. The finding that fluent readers deploy the left and right hand alternately for the spatial and verbal aspects of texts suggests that the functions can be divided equally, or almost equally, between the two hands, even for right-handed individuals. But the brief survey of findings suggested that shape or spatial coding cannot be inferred directly from left-hand performance without independent tests.

The behavioural test that was devised to assess spatial coding independently of hand effects showed that instructions to use body-centred cues, and cues from an external surrounding square, for reference improved recall accuracy for distance and location tasks very decisively. The effect was independent of movement differences and independent also of performance with the left and right hands. It was suggested that providing cues from two complementary sources of reference explicitly in spatial tasks would prove useful as an independent behavioural test of spatial coding.

EXTERNAL CUES CAN BE USED FOR REFERENCE IN HAPTIC TASKS

The new findings showed that external cues can be used for reference also with purely haptic sensory inputs, contrary to the notion that haptic inputs are necessarily tied to egocentric frames of reference and that allocentric coding is impossible without inputs from vision or other distal sensory cues.

To my knowledge, allocentric coding had not previously been demonstrated experimentally with purely haptic sensory inputs. Separating and combining the two types of cues showed that external cues produced the same level of accuracy of location recall when body-centred reference was disrupted as was found with intact body-centred cues in the absence of external reference. Moreover, the two forms of reference combined additively.

Modality systems certainly differ in the extent to which they routinely provide sensory cues from different sources. Haptic cues that are external to a target often have to be signalled explicitly, unless the space is familiar, so that such cues are known to be present and can be sought. But that link between modalities and forms of spatial coding depends on the information that is available in current task conditions and from longer-term experience. It is not an exclusive connection between modalities and forms of reference organisation. The point is that when external cues are made available, or explicit, in purely haptic conditions, they are as effective as body-centred cues for spatial reference.

The empirical evidence is consistent with the view that the processing activities that integrate diverse inputs for spatial reference are similar or in common between vision and touch. The form of spatial organisation that transpires is a function of the inputs that are available from different sources in given task conditions. The inputs that are being integrated for reference evidently preserve modality-specific aspects in perception. But the activities that are concerned with integrating the inputs spatially seem to follow the same rules in touch and vision.

Further studies are needed also on how different forms of reference relate to each other within the haptic modality. How, for instance, do conflicting versus congruent combinations of different types of potential reference cues affect haptic spatial coding compared to either touch or movement tasks alone? Most current studies on the relation between touch and vision have used vibrotactile stimulation, or passive touch. The approach here to questions about spatial processing has centred on active touch, or haptic perception, which depends on the combination of touch and exploratory or scanning movements. There are thus further question also about diverse spatial effects for active touch that remains stationary and for movements that exclude touch.

The findings that have been reviewed suggest that spatial processes embrace the whole gamut, from operating automatically, or without conscious awareness (Weiskrantz, 1986), in perception and performance, to procedures and cognitive skills that involve solving complex rotation, displacement and geometric problems. There is also a return to quasi-automatic, habitual procedures with sufficient experience and familiarity. A good deal of current neurophysiological and neuroimaging research is concerned with investigating the neocortical and subcortical neural circuits that are likely to be involved. Further studies are also needed on the effects of explicit instructions to use specified external and body-centred reference cues compared to conditions in which the same reference cues produce automatic effects in active touch and in vision.

COMMON FACTORS IN ILLUSIONS BY TOUCH AND VISION

Contrary to the view that the similarity of tactual illusions to visual illusions is fortuitous, three very different shape illusions could be explained by within-shape discrepancies and their relation to spatial reference information. The within-shape discrepancies, and their relation to other reference cues, differ between the illusions, but that also applies to the same illusions in vision.

An explanation of the visual vertical–horizontal illusion was that vertical lines are seen as longer because they occupy a larger proportion of the compressed vertical extent of the elliptical visual field, compared to the smaller proportion occupied by horizontal lines relative to the wider horizontal extent of the elliptical field. The same proportional rule applies to touch. Radial movements for vertical lines cover a large portion of the haptic field of possible movements that are anchored at the elbow joint. Horizontal lines cover a proportionately smaller extent of the much larger lateral sweep that is possible for movements that are anchored at the shoulder. Modality-specific factors contribute. The haptic vertical–horizontal illusion seems to be found only in some spatial planes. However, the similarity of the tactual to the visual illusion is explained by the relation of the constituent movement lengths to their respective spatial boundaries in touch as well as in vision.

The bisection illusion is as obvious in touch as it is in vision. It occurs in both modalities regardless of the orientation of the figure. The explanation was that the continuous bisecting line is felt as longer because the bisection point is perceived as a boundary that divides the bisected line into two shorter lengths. The shorter lengths make the continuous line look and feel longer. The explanation was supported in touch. The overestimation increased inversely with the length of line beyond the boundary point. Inspection of the visual figures suggested that the same explanation applies to the visual illusion, although that requires empirical confirmation.

The new evidence showed that the powerful Müller-Lyer illusions in vision and touch are virtually eliminated by the same instructions to use body-centred cues for reference. The instruction to ignore the fins was not sufficient alone to abolish the visual illusion either in the experiments described here or in previous visual studies. However, instructions to relate the perceptual inputs to body-centred cues produced the same steep decline of the illusion in vision as in touch. The finding suggests strongly that egocentric reference is a common factor in the integration of length stimuli from vision and touch, at least in that illusion.

The further questions concern the relation of discrepancies within shape features to external and body-centred reference information in shapes that do not produce the same illusion in vision and touch. Separate, rather than crossmodal, designs may be needed for the two modalities to avoid influences from the second modality.

Another area of future enquiry concerns the conditions in which body-centred and/or external cues are used spontaneously, as shown by better results with intact versus disrupted body-centred cues, in contrast to conditions in which such cues are present but only become effective with explicit instructions to use them. One possibility is that discrepancies in inputs inhibit normally habitual or automatic integrating activities.

It should be noted that the claim here is not that discrepancies in external, body-centred, or shape-centred reference cues are the only factors that lead to perceptual biases. Modality-specific aspects of inputs certainly have effects. The suggestion is, rather, that the spatial integration of inputs is an important factor in organising inputs from diverse sources. Integrating diverse inputs enters into the discrepancies, as well as into the convergence and overlap of inputs. Discrepant length inputs within shapes in different modalities produce biases that are in common between them. The convergence and overlap of spatially organised reference increases perceptual accuracy in both vision and touch.

VISION IMPROVES HAPTIC RECALL OF LOCATIONS ONLY IF IT AFFORDS ADDITIONAL REFERENCE CUES

The final prediction concerned the question of whether added vision has a general facilitating effect on inputs from touch and movement, even when it does not provide task-relevant information. Memory for locations in a tactile map was tested with and without full vision, and with three different forms of impaired vision that were simulated with specially designed spectacles.

Vision that excluded all spatial cues did not improve touch, in contrast to the other forms of reduced vision. It was tested here with spectacles that provided only diffuse light perception. Combining that with touch did not differ from touch alone either in overall accuracy or in the error pattern for different locations.

The two other forms of impaired vision contributed less to touch than full vision. But they also differed from each other in errors at two specific locations. The difference was explained by differences, respectively, in the clarity of focal vision that is preserved in tunnel vision, but not in peripheral vision, and the range of potential reference cues that is accessed more easily by using appropriate head movements in peripheral vision than it is in tunnel vision.

Further experiments on the combination of information from touch and different forms of residual vision are needed to extend our knowledge of the factors that operate in different forms of bimodal integration. Such studies are also needed on practical grounds. In principle, people with impaired vision can benefit from obtaining added information from haptic exploration. But the findings also suggest that, in the case of specific forms of visual impairments, we first need to establish the particular aspects of vision that should be supplemented and how they combine with information from touch and movement.

The results provided decisive evidence that an explicit distinction between spatial and non-spatial aspects of vision is needed in order to specify those aspects of vision that actually work in conjunction with touch and those that do not. Strong and converging evidence that spatial coding should be separated from vision was considered in the second chapter. The implication of the study that excluded spatial cues explicitly (Millar & Al-Attar, 2005) was that spatial and non-spatial aspects of vision need to be distinguished explicitly because they can have quite different effects in bimodal conditions.

Overall, the findings that have been reviewed here were consistent with the predictions from the theoretical description that prompted the tests. But the findings also suggested that the distinctness of features, unexpected local touch cues, and the importance of specific task conditions and materials need to be given greater weight.

CONCLUSIONS AND OUTLOOK

Three main assumptions were made in the theoretical description that I adopted originally. First, the metaphor of active processing in interrelated networks was used to describe the behavioural system.

Second, it was assumed that reference cues are crucial for spatial tasks. Spatial coding was defined as the activity of integrating converging inputs from diverse sources into reference cues that specify the locations, distances, or directions that a task demands.

The third emphasis was on the convergence or partial overlap of the inputs that could be organised spatially. It was assumed that the broad categories of reference organisations that emerge – whether body-centred and/or externally anchored, and/or shape-based – are determined by the inputs that are available in given task conditions and are influenced by longer-term experience and information.

Considering first the network metaphor that I originally adopted from neuroscience, it may be said that it no longer needs any defence. It is now the almost universally accepted description of brain and neural functions. A similar description is used in connectionist computational models that test rules that may describe behavioural and/or physiological relations mathematically.

The neurological picture has become considerably more detailed and complex, but also more clearly related to behavioural findings. Further refinements and improvements in the know-how of non-invasive techniques are likely to produce still greater advances in findings that correlate with the factors, and interrelations between factors, that behavioural experiments point up. If anything, the general description of the whole system in terms of converging active processing in interrelated networks fits even better with the new behavioural findings that were summarised here. The activities that I assume to characterise spatial coding can be attributed to neural circuits that involve a number of distributed, but interconnected, cortical and subcortical areas that also have a variety of subsidiary network connections.

There is little doubt that most spatial tasks also involve factors other than spatial coding. What other factors are involved does depend on the task, on the current information, and on the longer-term information that is available to the person. The current behavioural, as well as neurological, findings suggest "top-down", "bottom-up", and also lateral connections between the factors that are activated by the demands of spatial tasks.

The second and third points concerned what I have here called the "reference" hypothesis. The further findings seem to me to justify it as a useful, indeed a necessary, criterion for spatial coding. The importance of reference cues for spatial coding, and of the integration and need for overlapping stimuli, was shown by the study on veering in a space that totally lacked external reference cues that could be integrated with the heading direction. It was also shown by the significant increases in accuracy with instructions to use body-centred and/or external cues for reference in the recall of distances and locations. The organisation of length inputs with respect to their limitation by reference cues explained the tactual illusions by factors that also apply to same illusions in vision. Applying that rule virtually eliminated one of the strongest illusions in both vision and touch.

The differences in effects on touch from different types of impaired versus full or no vision showed the importance of distinguishing between spatial and non-spatial aspects of vision in explaining what aspects of vision affect touch in spatial tasks.

The summary of findings pointed to a number of future directions for work that will hopefully specify further interconnections between behavioural and neurological factors. The findings were consistent with the assumption that reference cues, which specify the location of an object, act as retrieval cues in memory tests. Many tasks involve immediate or very brief periods of short-term memory. A further direction of future work thus concerns delay conditions between initial presentation of reference cues and actions that depend on different types of retrieval cues in both large-scale and small-scale spaces.

The twentieth century saw changes from the firmly linear, and also from strictly "modular", theories of brain and behaviour (e.g. Fodor, 1983) to the current more flexible, interrelating network models. The changes have been fruitful. Progress in the foreseeable future is thus likely to come from further analyses of specific effects and their interconnections, rather than by a total change of approach, unless further studies that span gaps between disciplines suggest a different picture.

The sheer complexity, at the molecular and biochemical levels, of the self-organising activities that influence and are influenced by externally triggered changes, and their relation to neurological and psychological factors, makes it likely that the goal of describing the whole system is some way off. But it is a worthwhile goal.

So, where have we got to with the question of how the senses relate to ideas of space? I chose active touch as a most unpropitious spatial sense compared to vision and empirical studies on how spatial processing takes place as a means of finding answers.

Characterising spatial coding explicitly as an integrating activity has produced results that take us a step further, at least in showing that apparently contradictory findings and assumptions may both apply. A nice example is the finding that showed allocentric spatial coding with purely haptic inputs, and additive effects of allocentric and egocentric cues. It runs counter to the view that different forms of spatial coding are tied to particular sensory modalities. But the evidence does not fit in either with the notion of a higher-order, amodal system that applies equally to all modalities. Spatial coding, as an activity that integrates inputs from diverse sources, accommodates both types of evidence.

The findings that have been considered also suggest that some amendments to the original description are needed. The effect of the task that confronts an organism was included in the previous account. But the importance of tasks in eliciting specific combinations of activities probably needs to be stressed even more. The number and intricacy of neural connections that are being uncovered run counter to the metaphor of quite separate "modules" as a description of normal functions.

The factors that facilitate or constrain integrative processing were previously merely listed under the rubric of the amount, salience, and convergence of inputs from different sources. But a number of stimulus "characteristics" are in common between vision and touch and, indeed, between most sensory modalities. The characteristics are not "amodal". They are implemented very differently in different modalities. The proximity of stimuli relative to each other, and to the body of the observer, is one such characteristic. The impact of persistence of stimulation, in contrast to novelty or change in any stimulus, is another. The time, or the sequence, in which stimuli are seen or felt can also determine their impact on processing. The extent to which their impact coincides can account for the overlap or "redundancy" that increases accuracy in bimodal conditions. Discrepancies in stimulus characteristics that sensory inputs have "in common" explain how one or other modality may "win out" over the other in bimodal conditions. The rubric of "associative links" must here do service for the contexts in which the stimuli were experienced in the past, and the repetition or length of their combination with other stimuli.

A flowchart of some of the characteristics that seem to influence integrative activities (Figure 9.2) illustrates how additive versus disparate forms of reference organisation may arise.

The flowchart is not complete or exhaustive. The integrative activities of the very networks that play an essential role in response to the demands of spatial tasks are merely indicated in the chart. The chart highlights some shared, as well as diverse, aspects of the stimulation that specialised receptor systems analyse very differently. Both congruence and discrepancies in inputs influence the forms of spatial reference integration that organise perception and action. The integrating activities seem to follow common rules or principles that provide coherence to our ideas and representations of "space" or spaces. The inputs are specialised and differ. But the activities that integrate diverse inputs into spatial reference organisations for perception and action seem to follow the same probabilistic rules.

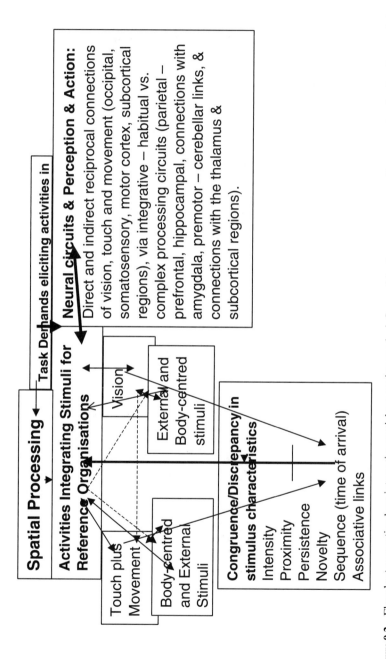

Figure 9.2 Flowchart suggesting how integrative spatial processing can be influenced by the congruence or discrepancy in stimulus characteristics that are in common between vision, touch and movement, although the characteristics are implemented differently in different modalities. (The important network connections for spatial processing in different tasks are here merely indicated in the "box" on the right-hand side.)

References

Adams, J. A., Bodis-Wollner, I., Enoch, J. M., Jeannerod, M., & Mitchell, D. E. (1990). Normal & abnormal mechanisms of vision: Visual disorders and visual deprivation. In L. Spillmann & J. S. Werner (Eds.), *Visual perception: The neurophysiological foundations* (pp. 381–416). New York: Academic Press.

Adams, J. A., & Dijkstra, S. (1966). Short-term memory for motor responses. *Journal of Experimental Psychology, 71*, 314–318.

Amedi, A., Jacobson, G., Hendler, T., Malach, R., & Zohary, E. (2002). Convergence of visual and tactile shape processing in the human lateral occipital complex. *Cerebral Cortex, 12*, 1202–1212.

Amedi, A., Malach, R., Hendler, T., Peled, S., & Zohary, E. (2001). Visuo-haptic object-related activation in the ventral visual pathway. *Nature Neuroscience, 4*, 324–330.

Andersen, R. A. (1999). Multimodal integration for the representation of space in the posterior parietal cortex. In N. Burgess, K. J. Jeffery, & J. O'Keefe (Eds.), *The hippocampal and parietal foundations of spatial cognition* (pp. 90–103). Oxford, UK: Oxford University Press.

Andersen, R. A., Snyder, L. H., Bradley, D. C., & Xing, J. (1997). Multimodal representation of space in the posterior parietal cortex and its use in planning movements. *Annual Review of Neuroscience, 20*, 303–330.

Andersen, R. A., & Zipser, D. (1988). The role of the posterior parietal cortex in coordinate transformations for visual-motor integration. *Canadian Journal of Physiology & Pharmacology, 66*, 488–501.

Anii, A., & Kudo, K. (1997). Effects of instruction and practice on the length-reproduction task using the Müller-Lyer figure. *Perceptual and Motor Skills, 85*, 819–825.

Appelle, S., & Countryman, M. (1986). Eliminating the haptic oblique effect: Influence of scanning incongruity and prior knowledge of the standard. *Perception, 15,* 325–329.

Appelle, S., & Gravetter, F. G. (1985). Effect of modality-specific experience and visual and haptic judgment of orientation. *Perception, 14,* 763–773.

Arbib, M. (1991). Interaction of multiple representations of space in the brain. In J. Paillard (Ed.), *Brain and space* (pp. 379–403). Oxford, UK: Oxford University Press.

Arieh, Y., & Marks, L. E. (2002). Context effects in visual length perception: Role of ocular, retinal and spatial location. *Perception & Psychophysics, 64,* 478–492.

Armstrong, L., & Marks, L. E. (1997). Differential effects of stimulus context on perceived length: Implications for the horizontal–vertical illusion. *Perception & Psychophysics, 59,* 1200–1213.

Armstrong, L., & Marks, L. E. (1999). Haptic perception of linear extent. *Perception & Psychophysics, 61,* 1211–1226.

Ashmead, D. H., Wall, R. S., Eaton, S. B., Ebinger, K. A., Snook-Hill, M. M., Guth, D. A., et al. (1998). Echolocation reconsidered: Using spatial variation in the ambient sound field to guide locomotion. *Journal of Visual Impairment and Blindness, 92,* 615–632.

Atkinson, J., Anker, S., Braddick, O., Nokes, L., & Mason, A. (2001). Visual and visuospatial development in young children with Williams syndrome. *Developmental Medicine & Child Neurology, 43,* 330–337.

Atkinson, J., Braddick, O., Anker, S., Curran, W. Andrew, R., Wattam, B., et al. (2003). Neurobiological models of visuo-spatial cognition in children with Williams syndrome: Measures of dorsal-steam and frontal function. *Developmental Neuropsychology, 23,* 139–172.

Atkinson, R. C., & Shiffrin, R. M. (1968). Human memory: A proposed system and its control processes. In K. Spence & J. T. Spence (Eds.), *The psychology of learning and motivation* (Vol. 2). London: Academic Press.

Avery, G. C., & Day, R. H. (1969). The basis of the horizontal–vertical illusion. *Journal of Experimental Psychology, 81,* 376–380.

Baddeley, A. D. (1986). *Working memory.* Oxford, UK: Clarendon Press.

Baddeley, A. D. (1990). *Human memory: Theory and practice.* Hove, UK: Lawrence Erlbaum Associates Ltd.

Baddeley, A. D. (2000). The episodic buffer: A new component of working memory? *Trends in Cognitive Science, 4,* 417–422.

Baddeley, A. D. (2007). *Working memory, thought, and action.* Oxford, UK: Oxford University Press.

Baddeley, A. D., & Hitch, G. J. (1974). Working memory. In G. H. Bower (Ed.), *The psychology of learning and motivation: Advances in research and theory* (Vol. 8). New York: Academic Press.

Ballesteros, S., Bardisa, D., Millar, S., & Reales, J. M. (2005). The haptic test battery: A new instrument to test tactual abilities in blind and visually impaired and sighted children. *British Journal of Visual Impairment and Blindness, 23,* 11–24.

Ballesteros, S., Millar, S., & Reales, S. (1998). Symmetry in haptic and in visual perception. *Perception & Psychophysics, 60,* 389–404.

REFERENCES

Barrett, D. J. K., Bradshaw, M. F., Rose, D., Everatt, P. J., & Simpson, P. J. (2001). Reflexive shifts of attention operate in an egocentric coordinate frame. *Perception, 30*, 1083–1091.

Bayliss, G. C., & Moore, B. O. (1994). Hippocampal lesions impair spatial response selection in the primate brain. *Experimental Brain Research, 98*, 110–118.

Bean, C. H. (1938). The blind have "optical illusions". *Journal of Experimental Psychology, 22*, 283–289.

Beers, R. J. van, Sittig, A., & Gon, J. J. (1999). Integration of proprioceptive and visual-position information: An experimentally supported model. *Journal of Neurophysiology, 81*, 1355–1364.

Beers, R. J. van, Wolpert, M., & Haggard, P. (2002). When feeling is more important than seeing in sensorimotor adaptation. *Current Biology, 12*, 834–837.

Behrmann, M. (2000). Spatial reference frames and hemispatial neglect. In M. S. Gazzaniga (Ed.), *The new cognitive neurosciences* (pp. 651–666). New York: MIT Press.

Berlá, E. P., & Butterfield, L. H. Jr. (1977). Tactual distinctive feature analysis: Training blind students in shape recognition and in locating shapes on a map. *Journal of Special Education, 11*, 336–346.

Bertelson, P., Mousty, P., & D'Alimonte, G. (1985). A study of braille reading: Patterns of hand activity in one-handed and two-handed reading. *Quarterly Journal of Experimental Psychology, 37A*, 235–256.

Berthoz, A. (1991). Reference frames for the perception and control of movement. In J. Paillard (Ed.), *Brain and space* (pp. 81–111). Oxford, UK: Oxford University Press.

Berthoz, A. (Ed.). (1993). *Multisensory control of movement*. Oxford, UK: Oxford University Press.

Bi, G.-G., & Poo, M.-M. (2001). Synaptic modification by correlated activity: Hebb's postulate revisited. *Annual Review of Neuroscience, 24*, 139–166.

Bisiach, E., & Vallar, G. (1988). Hemineglect in humans. In F. Boller & J. Grafman (Eds.), *Handbook of neurophysiology* (pp. 195–222). Amsterdam: Elsevier.

Bolles, R. C. (1969). The role of eye movements in the Müller-Lyer illusion. *Perception and Psychophysics, 6*, 175–176.

Bottini, G., Karnath, H. O., Vallar, G., Sterzi, R., Frith, C. D., Frackowiak, R. S. I., et al. (2001). Cerebral representation for egocentric space: Functional-anatomical evidence from caloric vestibular stimulation and neck vibration. *Brain, 124*, 1182–1196.

Boven, R. W. van, Hamilton, R. H., Kauffman, T., Keenan, J. P., & Pascual-Leone, A. (2000). Tactile spatial resolution in blind readers. *Neurology, 54*, 2230–2236.

Brabyn, J. A. (1978). *Laboratory studies of aided blind mobility*. PhD Thesis, University of Canterbury, New Zealand.

Brabyn, J. A., & Strelow, E. R. (1977). Computer-analyzed measures of characteristics of human locomotion and mobility. *Behavior Research Methods and Instrumentation, 9*, 456–462.

Bradshaw, J. L., & Nettleton, N. C. (1981). The nature of hemispheric specialization in man. *Behavioral & Brain Sciences, 4*, 51–91.

Bradshaw, J. L., Nettleton, N. C., & Spehr, K. (1982). Braille reading and left and right hemispace. *Neuropsychologia, 20*, 493–500.

Brainard, G. C., Rollag, M. D., & Hanifin, J. P. (1997). Photic regulation of melatonin in humans: Ocular and neural signal transduction. *Journal of Biological Rhythms*, *12*, 537–546.

Broadbent, D. E. (1958). *Perception and communication*. London: Pergamon Press.

Brooks, L. R. (1968). Spatial and verbal components of the act of recall. *Canadian Journal of Psychology*, *22*, 349–368.

Brown, J. A. (1958). Some tests of the decay theory of immediate memory. *Quarterly Journal of Experimental Psychology*, *10*, 12–21.

Bryant, P. E., Jones, P., Claxton, V., & Perkins, G. M. (1972). Recognition of shapes across modalities by infants. *Nature*, *240*, 303–304.

Büchel, C., Price, C., Frackowiak, R. S. J., & Friston, K. (1998). Different activation patterns in the visual cortex of late and congenitally blind subjects. *Brain*, *121*, 409–419.

Bürklen, K. (1932). *Touch reading of the blind* (F. K. Merry, Trans.). New York: American Foundation for the Blind.

Burnod, Y., & Dufosse, M. (1991). A model for the cooperation between cerebral cortex and cerebellar cortex in movement learning. In J. Paillard (Ed.), *Brain and space* (pp. 446–460). Oxford, UK: Oxford University Press.

Burton, H., Snyder, A. Z., Conturo, T. E., Akbudak, E. Ollinger, J. M., & Raichle, M. E. (2002). Adaptive changes in early and late blind: A fMRI study of Braille reading. *Journal of Neurophysiology*, *87*, 589–607.

Burtt, H. E. (1917). Tactual illusions of movement. *Journal of Experimental Psychology*, *2*, 371–385.

Calvert, G., Spence, C., & Stein, B. (Eds.). (2004). *Handbook of multisensory processes*. Cambridge, MA: MIT Press.

Carpenter, P. A., & Eisenberg, P. (1978). Mental rotation and the frame of reference in blind and sighted individuals. *Perception & Psychophysics*, *23*, 117–124.

Carrasco, M., Figueroa, J. G., & Willen, J. D. (1986). A test of the spatial frequency explanation of the Mueller-Lyer illusion. *Perception*, *15*, 553–562.

Cashdan, S. (1968). Visual and haptic form discrimination under conditions of successive stimulation. *Journal of Experimental Psychology*, *76*, 215–218.

Casla, M., Blanco, F., & Travieso, D. (1999). Haptic perception of geometric illusions by persons who are congenitally totally blind. *Journal of Visual Impairment and Blindness*, *93*, 583–588.

Castillo, M., & Butterworth, G. (1981). Neonatal localisation of a sound in visual space. *Perception*, *10*, 331–338.

Chapman, E. K., Tobin, M. J., Tooze, S., & Moss, S. (1989). *Look and think: A handbook for teachers*. London: R.N.I.B.

Cheng, M. F. H. (1968). Tactile–kinaesthetic perception of length. *American Journal of Psychology*, *81*, 74–82.

Cho, Y. S., & Proctor, R. W. (2005). Representing response positions relative to display location: Influence of orthogonal stimulus–response compatibility. *Quarterly Journal of Experimental Psychology*, *58A*, 839–864.

Chow, K. L., Riesen, A. H., & Newell, F. W. (1957). Degeneration of ganglion cells in infant chimpanzees reared in darkness. *Journal of Comparative Neurology*, *107*, 27–42.

Claparède, E. (1943). L'orientation lointaine. In *Nouveau traité de psychologie* (Vol. 8). Paris: Presses Universitaires de France.

Cohen, L., Celnik, P., Pascual-Leone, A., Corwell, B., Faiz, L., Dambrosia, J., et al. (1997). Functional relevance of cross-modal plasticity in blind humans. *Nature, 389*, 180–183.

Collani, G. von (1985). Retinal projection or size constancy as determinants of the horizontal–vertical illusion. *Perceptual & Motor Skills, 61*, 547–557.

Connolly, K., & Jones, B. (1970). A developmental study of afferent–efferent integration. *British Journal of Psychology, 61*, 259–266.

Conrad, R. (1964). Acoustic confusions in immediate memory. *British Journal of Psychology, 55*, 75–84.

Conrad, R. (1971). The chronology of the development of covert speech in children. *Developmental Psychology, 5*, 398–405.

Coren, S. (1970). Lateral inhibition and geometric illusions. *Quarterly Journal of Experimental Psychology, 22*, 274–278.

Coren, S., Girgus, J. S., & Schiano, D. (1986). Is adaptation of orientation-specific cells a plausible explanation of illusion decrement? *Bulletin of the Psychonomic Society, 24*, 207–210.

Coren, S., & Porac, C. (1983). The creation and reversal of the Müller-Lyer illusion through attentional modulation. *Perception, 12*, 49–54.

Cornoldi, C., & Vecchi, T. (2003). *Visuo-spatial working memory and individual differences*. Hove, UK: Psychology Press.

Craig, J. C., & Johnson, K. O. (2000). The two-point threshold: Not a measure of tactile spatial resolution. *Current Directions in Psychological Science, 9*, 29–32.

Cratty, B. J. (1967). The perception of gradient and the veering tendency while walking without vision. *American Foundation for the Blind Research Bulletin, 14*, 31–51.

Cratty, B. J. (1971). *Movement and spatial awareness in blind children and youth*. Springfield, IL: Charles Thomas.

Cratty, B. J., & Williams, M. (1966). *Perceptual thresholds of non-visual locomotion* (Part 2). Los Angeles: University of California.

Cuijpers, R. H., Kappers, A. M. L., & Koenderlink, J. J. (2002). Visual perception of co-linearity. *Perception & Psychophysics, 64*, 352–356.

Danziger, S., Kingstone, A., & Wards, R. (2001). Environmentally defined frames of reference: Their time course and sensitivity to spatial cues and attention. *Journal of Experimental Psychology: Human Perception and Performance, 27*, 494–503.

Darwin, C. (1879). *On the origin of the species by means of natural selection*. London: Murray.

Davenport, R. K., & Rogers, C. M. (1970). Intermodal equivalence of stimulation in apes. *Science, 168*, 279–280.

Davidson, P. W. (1972). The role of exploratory activity in haptic perception: Some issues, data and hypotheses. *Research Bulletin, American Foundation for the Blind, 24*, 21–28.

Davidson, P. W., Appelle, S., & Haber, R. N. (1992). Haptic scanning of braille cells by low and high proficiency braille readers. *Research in Developmental Disabilities, 13*, 99–111.

Davidson, P. W., Wiles-Kettleman, M., & Haber, R. N. (1980). Relationship between handmovements, reading competence and passage difficulty in Braille reading. *Neuropsychologia, 18*, 629–635.

Davidon, R. S., & Cheng, M. F. H. (1964). Apparent distance in horizontal tactile kinesthetic stimuli. *Quarterly Journal of Experimental Psychology, 19*, 74–78.

Day, R. H. (1962). The effects of repeated trials and prolonged fixation on error in the Müller-Lyer figure. *Psychological Monographs, 76*, Whole No: 533.

Day, R. H., & Avery, G. C. (1970). Absence of the horizontal–vertical illusion in haptic space. *Journal of Experimental Psychology, 83*, 172–173.

Day, R. H., & Wong, T. S. (1971). Radial and tangential movement directions as determinants of the haptic illusion in an L figure. *Journal of Experimental Psychology, 87*, 19–22.

Delantionio, A., & Riggio, L. (1981). Hemisphere lateralization and tactile modality. *Ricerche di Psicologia, 5*, 275–310.

Della Sala, S., Gray, C., Baddeley, A., & Wilson, L. (1999). Pattern span: A tool for unwelding visuo-spatial memory. *Neuropsychologia, 37*, 1189–1199.

DeLucia, P., Longmire, S. P., & Kennish, J. (1994). Diamond-winged variants of the Mueller-Lyer figure: A test of Virsu's centroid theory. *Perception & Psychophysics, 55*, 287–295.

Deregowski, J., & Ellis, H. D. (1972). Effect of stimulus orientation upon haptic perception of the horizontal–vertical illusion. *Journal of Experimental Psychology, 95*, 14–19.

De Renzi, E. (1978). Hemisphere asymmetry as evidenced by spatial disorders. In M. Kinsbourne (Ed.), *Asymmetrical functions of the brain*. Cambridge, UK: Cambridge University Press.

De Renzi, E. (1982). *Disorders of space exploration and cognition*. New York: John Wiley.

De Renzi, E., Faglioni, P., Lodesani, M., & Vecchi, A. (1983). Performance of left brain-damaged patients on imitation of single movements and motor sequences: Frontal and parietal patients compared. *Cortex, 19*, 333–343.

De Renzi, E., Faglioni, P., & Scotti, G. (1969). Impairment of memory for position following brain damage. *Cortex, 5*, 274–284.

Dewar, R. E. (1967). Stimulus determinants of the practice decrement of the Müller-Lyer illusion. *Canadian Journal of Psychology, 21*, 504–520.

Dodd, B., & Campbell, R. (Eds.). (1987). *Hearing by eye: The psychology of lip-reading*. Hove, UK: Lawrence Erlbaum Associates Ltd.

Duhamel, J.-R., Colby, C. L., & Goldberg, M. E. (1991). Congruent representations of visual somatosensory space in single neurons of the monkey ventral intraparietal cortex (Area VIP). In J. Paillard (Ed.), *Brain and space* (pp. 223–236). Oxford, UK: Oxford University Press.

Ebert, P. C., & Pollack, R. H. (1972). Magnitude of the Mueller-Lyer illusion as a function of hue, saturation and fundus pigmentation. *Psychonomic Science, 26*, 225–226.

Erlebacher, A., & Sekuler, R. (1969). A conclusion on confusion in the illusion of Müller-Lyer. *Proceedings of the Annual Convention of the American Psychological Association, 4*, 27–28.

Ernst, M. O., & Banks, M. S. (2002). Humans integrate visual and haptic information in a statistically optimal fashion. *Nature, 415*, 429–433.

Ernst, M. O., Banks, M. S., & Buelthoff, H. H. (2000). Touch can change visual slant perception. *Nature Neuroscience, 3*, 69–73.

Ettlinger, G. (1967). Analysis of cross-modal effects and their relationship to language. In F. L. Darley & C. H. Millikan (Eds.), *Brain mechanisms underlying speech and language*. New York: Grune & Stratton.

Evans, G. (1985). *Collected papers*. Oxford, UK: Oxford Clarendon Press.

Farrell, M. J., & Thomson, J. A. (1998). Automatic spatial updating during locomotion without vision. *Quarterly Journal of Experimental Psychology, 51A*, 637–654.

Fellows, B. J. (1967). Reversal of the Müller-Lyer illusion with changes in the inter-fins line. *Quarterly Journal of Experimental Psychology, 19*, 208–214.

Fellows, B. J. (1973). Comments on assimilation theory and the reversed Müller-Lyer illusion. *Perception, 2*, 219–223.

Fertsch, P. (1947). Hand dominance in reading braille. *American Journal of Psychology, 60*, 335–349.

Fessard, A. (1961). The role of neuronal networks in sensory communication within the brain. In W. A. Rosenblith (Ed.), *Sensory communication*. Cambridge, MA: MIT Press.

Festinger, L., White, C. W., & Allyn, M. R. (1968). Eye-movements and decrement in the Müller-Lyer illusion. *Perception & Psychophysics, 3*, 376–382.

Fick, A. (1851). *De errore quodam optico asymmetrica bulbi effecto*. Marburg: J. A. Kochii.

Fine, I., Wade, A. R., Brewer, A. A., May, M. G., Goodman, D., Boynton, G. M., et al. (2003). Long-term deprivation affects visual perception and cortex. *Nature Neuroscience, 6*, 915–916.

Finger, F. W., & Spelt, D. K. (1947). The illustration of the horizontal–vertical illusion. *Journal of Experimental Psychology, 37*, 243–250.

Fisher, G. H. (1970). An experimental and theoretical appraisal of the perspective and size-constancy theories of illusions. *Quarterly Journal of Experimental Psychology, 22*, 631–652.

Fitts, P. M., & Seeger, C. M. (1953). S-R compatibility: Spatial characteristics of stimulus and response codes. *Journal of Experimental Psychology, 46*, 199–210.

Fodor, J. (1983). *The modularity of mind: An essay on faculty psychology*. Cambridge, MA: MIT Press.

Foulke, E. (1982). Reading Braille. In W. Schiff & E. Foulke (Eds.), *Tactual perception: A source book*. New York: Cambridge University Press.

Fraisse, P. (1971). Temporal integration of elements in optical-geometric illusions and inversion of the Müller-Lyer Illusion. *Année Psychologique, 71*, 53–72.

Frisby, J., & Davis, I. (1971). Is the haptic Müller-Lyer illusion a visual phenomenon? *Nature, 231*, 463–465.

Fry, C. L. (1975). Tactual illusions. *Perceptual & Motor Skills, 40*, 955–960.

Galati, G., Committeri., G., Sanes, J. N., & Pizzamiglio, L. (2001). Spatial coding of visual and somatic sensory information in body-centred coordinates. *European Journal of Neuroscience, 14*, 737–746.

Galati, G., Lobel, E., Vallar, G., Berthoz, A., Pizzamiglio, L., & LeBihan, D. (2000). A neural basis of egocentric and allocentric coding of space in humans: A functional magnetic resonance study. *Experimental Brain Research, 133*, 156–164.

Gazzaniga, M. S. (1988). The dynamics of cerebral specialization and modular interactions. In L. Weiskrantz (Ed.), *Thought without language.* Oxford, UK: Clarendon Press.

Gazzaniga, M. S. (Ed.). (1994). *The cognitive neurosciences.* Cambridge, MA: MIT Press.

Gazzaniga, M. S., & LeDoux, J. E. (1978). *The integrated mind.* New York: Plenum Press.

Gentaz, E., & Hatwell, Y. (1996). Role of gravitational cues in the haptic perception of orientations. *Perception & Psychophysics, 56*, 1278–1292.

Gentaz, E., & Hatwell, Y. (1998). The haptic oblique effect in the perception of rod orientation in blind adults. *Perception & Psychophysics, 60*, 157–167.

Gentaz, E., & Hatwell, Y. (2000). Haptic processing of spatial and material object properties. In Y. Hatwell, A. Streri, & E. Gentaz (Eds.), *Touching for knowing: Cognitive psychology of haptic manual perception* (pp. 123–159). Amsterdam & Philadelphia: John Benjamins.

Gentaz, E., & Hatwell, Y. (2004). Geometric haptic illusions: Role of exploratory movements in the Müller-Lyer, Vertical–Horizontal and Delboeuf illusions. *Psychonomic Bulletin & Review, 11*, 31–40.

Gentilucci, M., Chieffi, S., Daprati, E., Saetti, M. C., & Toni, I. (1996). Visual illusion and action. *Neuropsychologia, 34*, 369–376.

Gescheider, G. A. (1997). *Psychophysics: The fundamentals.* Mahwah, NJ: Lawrence Erlbaum Associates Inc.

Gibson, E. J. (1969). *Perceptual learning and development.* New York: Appleton-Century-Crofts.

Gibson, J. J. (1962). Observations on active touch. *Psychological Review, 69*, 477–491.

Gibson, J. J. (1966). *The senses considered as perceptual systems.* Boston, MA: Houghton Mifflin.

Gibson, J. J. (1979). *The ecological approach to visual perception.* Boston, MA: Houghton Mifflin.

Gillan, D. J., Schmidt, W., & Hanowski, R. J. (1999). The effect of the Müller-Lyer illusion on map reading. *Perception & Psychophysics, 61*, 1154–1167.

Girgus, J., & Coren, S. (1973). Stability of forms under conditions of illusory distortion. *Perceptual & Motor Skills, 37*, 715–719.

Glennerster, A., & Rogers, B. (1993). New depth to the Müller-Lyer illusion. *Perception, 22*, 691–704.

Goldenberg, G., Podreka, I., & Steiner, M. (1990). The cerebral localization of visual imagery: Evidence from emission computerized tomography of cerebral blood flow. In P. J. Hampson, D. Marks, & J. T. E. Richardson (Eds.), *Imagery: Current developments.* London: Routledge.

Goldman-Rakic, P. S., & Friedman, H. R. (1991). The circuitry of working memory revealed by anatomy and metabolic imaging. In H. S. Levin, H. M. Eisenberg, & A. L. Benton (Eds.), *Frontal lobe function and dysfunction.* Oxford, UK: Oxford University press.

Goldreich, D., & Kanics I. M. (2003). Tactile acuity is enhanced in blindness. *Journal of Neuroscience, 23*, 3439.
Gollin, E. S. (1960). Developmental studies of visual recognition of incomplete objects. *Perceptual & Motor Skills, 11*, 289–298.
Goodnow, J. J. (1971). Eye and hand: Differential memory and its effect on matching. *Neuropychologica, 42*, 1187–1201.
Goryo, K., Robinson, J. O., & Wilson, J. A. (1984). Selective looking and the Müller-Lyer illusion: The effect of changes in the focus of attention on the Müller-Lyer illusion. *Perception, 13*, 647–654.
Grant, A. C., Thiagarajah, M. C., & Sathian, K. (2000). Tactile perception in blind Braille readers. *Perception & Psychophysics, 62*, 301–312.
Graziano, M. S., Cooke, F. D., & Taylor, C. S. R. (2000). Coding the location of the arm by sight. *Science, 290*, 1782–1786.
Graziano, M. S. A., & Gross, C. G. (1993). A bimodal map of space: Somatosensory receptive fields in the macaque putamen with corresponding visual receptive fields. *Experimental Brain Research, 97*, 96–109.
Graziano, M. S., & Gross, C. G. (1994). The representation of extrapersonal space: A possible role for bimodal visual–tactile neurons. In M. Gazzaniga (Ed.), *The cognitive neurosciences* (pp. 1021–1034). Cambridge, MA: MIT Press.
Green, D. M., & Swets, J. A. (1966). *Signal detection theory and psychophysics.* New York: John Wiley.
Gregory, R. L. (1963). Distortion of visual space as inappropriate constancy scaling. *Nature, 199*, 678–680.
Gregory, R. L. (1966). *Eye and brain: The psychology of seeing.* London: Weidenfeld & Nicolson.
Gregory, R. L. (1990). *Eye and brain: The psychology of seeing* (4th ed.). Oxford, UK: Oxford University Press.
Gregory, R. L. (1998). *Eye and brain: The psychology of seeing* (5th ed.). Oxford, UK: Oxford University Press.
Gregory, R. L. (2003). Seeing after blindness. *Nature Neuroscience, 6*, 909–910.
Gregory, R. L., & Wallace, J. G. (1963). *Recovery from early blindness: A case study. Experimental Psychological Society Monograph, No. 2.* Cambridge, UK: Heffer.
Griggs, R. (1974). Constancy scaling and the Mueller-Lyer illusion: More disconfirming evidence. *Bulletin of the Psychonomic Society, 4*, 168–170.
Gross, C. G., & Graziano, M. S. A. (1995). Multiple representations of space in the brain. *The Neuroscientist, 1*, 43–50.
Grunewald, A. P. (1966). A braille reading machine. *Science, 154*, 144–146.
Guest, S., & Spence, C. (2003). Tactile dominance in speeded discrimination of textures. *Experimental Brain Research, 150*, 201–207.
Gugerty, L., & Brooks, J. (2001). Seeing where you are heading: Integrating environmental and egocentric reference frames in cardinal direction judgments. *Journal of Experimental Psychology: Applied, 7*, 251–266.
Guldberg, G. A. (1897). Die Circularbewegung als thierische Grundbewegung, ihre Ursache Phänomenalität und Bedeutung. *Zeitschrift für Biologie, 35*, 419–458.
Gurfinkel, V. S., & Levick, Y. S. (1991). Perceptual and automatic aspects of the

postural body scheme. In J. Paillard (Ed.), *Brain and space* (pp. 147–162). Oxford, UK: Oxford University Press.

Guth, D., & LaDuke, R. (1994). The veering tendency of blind pedestrians: An analysis of the problem and literature review. *Journal of Visual Impairment and Blindness*, *88*, 391–400.

Guth, D., & LaDuke, R. (1995). Veering by blind pedestrians: Individual differences and their implications for instruction. *Journal of Visual Impairment and Blindness*, *89*, 28–37.

Haggard, P., Newman, C., Blundell, J., & Andrew, H. (2000). The perceived position of hand in space. *Perception & Psychophysics*, *62*, 363–377.

Harris, J. C. (1967). Veering tendency as a function of anxiety in the blind. *American Foundation for the Blind Bulletin*, *14*, 53–63.

Hatwell, Y. (1960). Étude de quelque illusions géométriques tactile chez les aveugles. *L'Année Psychologique*, *60*, 11–27.

Hebb, D. O. (1949). *The organization of behavior*. New York: John Wiley.

Held, R. (1963). Plasticity in human sensori-motor control. *Science*, *142*, 455–462.

Held, R. (1965). Plasticity in sensory-motor systems. *Scientific American*, *213*, 84–94.

Held, R., & Bauer, J. A. (1967). Visually guided reaching in infant monkeys after restricted rearing. *Science*, *155*, 718–720.

Held, R., & Hein, A. (1963). Movement produced stimulation in the development of visually guided behavior. *Journal of Comparative Physiology and Psychology*, *56*, 872–876.

Heller, M. A. (1983). Active and passive tactual recognition of form. *Journal of General Psychology*, *108*, 225–229.

Heller, M. A. (1984a). Active and passive tactile braille recognition. *Bulletin of the Psychonomic Society*, *24*, 201–202.

Heller, M. A. (1984b). Active and passive touch: The influence of exploration time on form recognition. *Journal of General Psychology*, *110*, 243–249.

Heller, M. A. (1989). Texture perception in sighted and blind observers. *Perception & Psychophysics*, *45*, 49–54.

Heller, M. A., Brackett, D. D., Salik, S. S., & Scroggs, E. (2003). Objects, raised lines and the haptic horizontal–vertical illusion. *Quarterly Journal of Experimental Psychology*, *56A*, 891–907.

Heller, M. A., Brackett, D. D., Wilson, K., Yoneyama, K., Boyer, A., & Steffen, H. (2002). The haptic Müller-Lyer illusion in sighted and blind people. *Perception*, *31*, 1263–1274.

Heller, M. A., Calcaterra, J. A., Burson, L. L., & Green, S. L. (1997). The tactual horizontal–vertical illusion depends on radial motion of the entire arm. *Perception & Psychophysics, 59*.

Heller, M. A., Calcaterra, J. A., Green, S. L., & Barnette, S. L. (1999). Perception of the horizontal and vertical in tangible displays: Minimal gender differences. *Perception*, *28*, 387–394.

Heller, M. A., Calcaterra, J. A., Green, S. L., & de Lima, F. (1999). The effect of orientation on Braille recognition in persons who are sighted and blind. *Journal of Visual Impairment & Blindness*, *93*, 416–418.

Heller, M. A., & Joyner, T. D. (1993). Mechanisms in the haptic horizontal–vertical illusion: Evidence from sighted and blind subjects. *Perception & Psychophysics, 53,* 422–428.

Heller, M. A., Joyner, T. D., & Dan-Fodio, H. (1993). Laterality effects in the haptic horizontal–vertical illusion. *Bulletin of the Psychonomic Society, 31,* 440–442.

Heller, M. A., & Myers, D. S. (1983). Active and passive tactual recognition of form. *Journal of General Psychology, 108,* 225–229.

Helmholtz, H. von (1867). *Handbuch der physiologischen Optik.* Leipzig: L. Voss.

Helmholtz, H. von (1896). *Handbuch der physiologischen Optik* (Vol. 3). Leipzig: L. Voss.

Henry, L., & Millar, S. (1991). Memory span increase with age: A test of two hypotheses. *Journal of Experimental Child Psychology, 51,* 459–484.

Henry, L., & Millar, S. (1993). Why does memory span improve with age? A review of the evidence for two current hypotheses. *European Journal of Cognitive Psychology, 5,* 241–287.

Hermelin, B., & O'Connor, N. (1971). Functional asymmetry in the reading of braille. *Neuropsychologica, 9,* 431–435.

Holdstock, J. S., Mayes, A. R., Cezayirli, C. L., Isaak, J. P., Aggleton, J. P., & Roberts, N. (2000). A comparison of egocentric and allocentric spatial memory in a patient with selective hippocampal damage. *Neuropsychologia, 38,* 410–425.

Hollins, M. (1986). Mental haptic rotation: More consistent in blind subjects? *Journal of Visual Impairment and Blindness, 80,* 950–952.

Hollins, M. (1989). *Understanding blindness.* Hillsdale, NJ: Lawrence Erlbaum Associates Inc.

Hollins, M., & Goble, A. K. (1988). Perception of the length of voluntary movements. *Somatosensory Research, 5,* 335–348.

Hollins, M., & Kelley, E. K. (1988). Spatial updating in blind and sighted people. *Perception & Psychophysics, 43,* 380–388.

Howard, I. P., & Rogers, B. J. (1995). *Binocular vision and stereopsis.* Oxford, UK: Oxford University Press.

Howard, I. P., & Templeton, W. B. (1966). *Human spatial orientation.* New York: John Wiley.

Hummel, F. C., & Cohen, L. G. (2005). Drivers of brain plasticity. *Current Opinion in Neurology, 18,* 667–674.

Huttenlocher, P. R. (2002). *Neural plasticity: The effects of the environment on the development of the cerebral cortex.* Cambridge, MA: Harvard University Press.

Hyvärinen, J., Carlson, S., & Hyvärinen, L. (1981). Early visual deprivation alters modality of neuronal responses in area 19 of the monkey cortex. *Neuroscience Letters, 26,* 239.

Hyvärinen, J., & Poranen, A. (1978). Receptive field integration and submodality convergence in the hand area of the post-central gyrus of the alert monkey. *Journal of Physiology, 283,* 539–556.

Igel, A., & Harvey, L. O. Jr. (1991). Spatial distortions in visual perception. *Gestalt Theory, 13,* 210–231.

James, T. W., Humphrey, G. K., Gati, J. S., Menon, R. S., & Goodale, M. A. (2002).

Differential effects of viewpoint on object-driven activation of dorsal and ventral streams. *Neuron, 35,* 793–801.

Jastrow, J. (1886). The perception of space by disparate senses. *Mind, 11,* 539–554.

Jeannerod, M. (1988). *The neural and behavioural organization of goal directed movements.* Oxford, UK: Clarendon Press.

Jones, B., & Connolly, K. (1970). Memory effects in cross-modal matching. *British Journal of Psychology, 61,* 267–270.

Jones, D., Farrand, P., Stuart, G., & Morris, N. (1995). The functional equivalence of verbal and spatial information in serial short-term memory. *Journal of Experimental Psychology: Learning, Memory, and Cognition, 21,* 1008–1018.

Jouen, F. (1992). Head position and posture in newborn infants. In A. Berthoz, W. Graf, & P. P. Vidal (Eds.), *The head–neck sensory motor system* (pp. 118–120). Oxford, UK: Oxford University Press.

Judd, C. H. (1899). A study of geometric illusions. *Psychological Review, 6,* 241–261.

Judd, C. H. (1902). Practice and its effects on the perception of illusions. *Psychological Review, 9,* 27–39.

Judd, C. H. (1905). The Müller-Lyer illusion. *Psychological Review, 7,* 55–81.

Kant, I. (1781). *Kritik der Reinen Vernunft* (R. Schmidt, Ed.). Hamburg: Felix Meiner Verlag, 1956.

Kappers, A. M. L. (2002). Haptic perception of parallelity in the midsagittal plane. *Acta Psychologica, 109,* 25–40.

Kappers, A. M. L., & Koenderink, J. J. (1999). Haptic perception of spatial relations. *Perception, 28,* 781–796.

Katz, D. (1925). *Der Aufbau der Tastwelt.* Leipzig: Barth.

Katz, D. (1925/1989). *The world of touch* (L. E. Krueger, Ed. & Trans). Hillsdale, NJ: Lawrence Erlbaum Associates Inc.

Katz, L. C., & Shatz, C. J. (1996). Synaptic activity and the construction of cortical circuits. *Science, 274,* 1133–1138.

Kennett, S., Taylor-Clarke, M., & Haggard, P. (2001). Noninformative vision improves spatial resolution of touch in humans. *Current Biology, 11,* 1188–1191.

Keulen, R. F., Adam, J. J., Fischer, M. H., Kuipers, H., & Jolles, J. (2002). Selective reaching: Evidence for multiple frames of reference. *Journal of Experimental Psychology: Human Perception and Performance, 28,* 515–526.

Kimura, D. (1973). The asymmetry of the human brain. *Scientific American, 228,* 70–78.

Kimura, D. (1993). *Neuromotor mechanisms in human communication.* New York: Oxford University Press.

Kimura, D., & Dunford, M. (1974). Normal studies on the function of the right hemisphere. In J. G. Dimond & S. J. Beaumont (Eds.), *Hemisphere function in the human brain.* London: Elek Sciences.

Klatzky, R. L. (1999). Path completion after haptic exploration without vision. *Perception & Psychophysics, 61,* 220–235.

Klatzky, R. L., Lederman, S. J., & Reed, C. L. (1989). Haptic integration of object properties: Texture, hardness, and planar contour. *Journal of Experimental Psychology: Human Perception and Performance, 15,* 45–57.

Klatzky, R. L., Loomis, J. M., Gollege, R. G., Cicinelli, J. G., et al. (1990). Acquisition

of route and survery knowledge in the absence of vision. *Journal of Motor Behavior, 22*, 19–43.
Klauer, K. C., & Zhao, Z. (2004). Double dissociations in visual and spatial short-term memory. *Journal of Experimental Psychology, 133*, 355–381.
Kobor, I., Furedi, L., Kovacs, G., & Spence, C. (2006). Back-to-front: Improved tactile discrimination performance in the space you cannot see. *Neuroscience Letters, 400*, 163–167.
Koffka, K. (1935). *Principles of Gestalt psychology*. New York, Harcourt & Brace.
Kosslyn, S. M. (1980). *Image and mind*. Harvard, MA: Harvard University Press.
Kosslyn, S. M. (1981). The medium and the message in mental imagery: A theory. *Psychological Review, 88*, 46–66.
Kosslyn, S. M. (1987). Seeing and imaging in the two hemispheres. *Psychological Review, 94*, 148–75.
Kosslyn, S. M. (1994). *Image and brain*. Cambridge, MA: MIT Press.
Krauthammer, G. (1968). Form perception across sensory modalities. *Neuropsychologica, 6*, 105–113.
Kress, G., & Cross, J. (1969). Visual and tactual interaction in judgments of the vertical. *Psychonomic Science, 14*, 165–166.
Krueger, L. E. (1973). Effects of irrelevant surrounding material on speed of same–different judgments of two adjacent stimuli. *Journal of Experimental Psychology, 98*, 252–259.
Kujala, T., Alho, K., Huotilainen, M., Ilmoniemi, R. J., Lehtokoski, A., et al. (1997). Electrophysiological evidence for cross-modal plasticity in humans with early-and late-onset blindness. *Psychophysiology, 34*, 213–216.
Kujala, T., Alho, K., Paavilainen, P., Summala, H., & Naatanen, R. (1992). Neural plasticity in processing of sound location by the early blind: An event-related potential study. *Electroencephalography and Clinical Neurophysiology, 84*, 469–472.
Kujala, T., Huotilainen, M., Sinkonen, J., Ahonen, A. I., Alho, K., Hamalaien, M. S., et al. (1995). Visual cortex activation during sound discrimination. *Neuroscience Letters, 183*, 143–146.
Kumar, S. (1977). Short-term memory for a non-verbal tactual task after cerebral commissurotomy. *Cortex, 13*, 55–61.
Künnapas, T. M. (1955a). An analysis of the "vertical–horizontal illusion". *Journal of Experimental Psychology, 49*, 134–140.
Künnapas, T. M. (1955b). Influence of frame size on apparent length of a line. *Journal of Experimental Psychology, 50*, 168–170.
Künnapas, T. M. (1957a). The vertical–horizontal illusion and the visual field. *Journal of Experimental Psychology, 53*, 405–407.
Künnapas, T. M. (1957b). Interocular differences in the vertical–horizontal illusion. *Acta Psychologica, 13*, 253–259.
Künnapas, T. M. (1957c). Vertical–horizontal illusion and surrounding field. *Acta Psychologica, 13*, 35–42.
Künnapas, T. M. (1958). Influence of head inclination on the vertical–horizontal illusion. *Journal of Psychology, 46*, 179–185.
Künnapas, T. M. (1959). The vertical–horizontal illusion in artificial visual fields. *Journal of Psychology, 47*, 41–48.

Kusajima, T. (1970). *Experimentelle Untersuchungen zum Augenlesen und Tastlesen.* Neuburgweiser: G. Schindele Verlag.

Laabs, G. L. (1973). Retention characteristics of different reproduction cues in short-term memory. *Journal of Experimental Psychology, 100,* 168–177.

Laabs, G. J., & Simmons, R. W. (1981). Motor memory. In D. Holding (Ed.), *Human skills.* New York: Wiley.

Ladavas, E. (1987). Influence of handedness on spatial compatibility effects with perpendicular arrangements of stimuli and responses. *Acta Psychologica, 64,* 13–23.

Langdon, D., & Warrington, E. K. (2000). The role of the left hemisphere in spatial processing. *Cortex, 36,* 691–702.

Lawrence, B. M., Myerson, J., Oonk, H. M., & Abrams, R. A. (2001). The effects of eye and limb movements on working memory. *Memory, 9,* 433–444.

Lechelt, E. C. (1982). The stability of tactile laterality differences across manipulations of temporal and spatial stimulus patterning. *International Journal of Rehabilitation Research, 5,* 370–372.

Lechelt, E. C., & Verenka, A. (1980). Spatial anisotropy in intra-modal and cross-modal judgements of stimulus orientation: The stability of the oblique effect. *Perception, 9,* 581–589.

Lederman, S. J., Klatzky, R. L., & Barber, P. O. (1985). Spatial and movement based heuristics for encoding pattern information through touch. *Journal of Experimental Psychology: General, 114,* 33–49.

Lederman, S. J., Klatzky, R. L., Chataway, C., & Summers, C. D. (1990). Visual mediation and the haptic recognition of two-dimensional pictures of common objects. *Perception & Psychophysics, 47,* 54–64.

Leinonen. L., Hyvärinen, J., Nyman, G., & Linnankoski, I. (1979). Functional properties of neurons, in lateral part of associative area 7 in awake monkeys. *Experimental Brain Research, 34,* 299–320.

Leonard, J. A., & Newman, R. C. (1967). Spatial orientation in the blind. *Nature, 215,* 1413–1414.

Lessard, N., Paré, M., Lepore, F., & Lassonde, M. (1998). Early-blind human subjects localize sound sources better than sighted subjects. *Nature, 395,* 278–280.

Levine, D. S. (2002). Neural network modeling. In H. Pashler & J. Wixted (Eds.), *Stevens' handbook of experimental psychology* (Vol. 4, pp. 223–269). New York: Wiley.

Lewald, J. (2007). More accurate sound localization induced by short-term light deprivation. *Neuropsychologia, 45,* 1215–1222.

Liben, L. S. (1988). Conceptual issues in the development of spatial cognition. In J. Stiles-Davis, M. Kritchevsky, & U. Bellugi (Ed.), *Spatial cognition.* Hillsdale, NJ: Lawrence Erlbaum Associates Inc.

Liotti, M., Ryder, K., & Woldorff, M. G. (1998). Auditory attention in the congenitally blind. Where and what gets reorganized? *Neuroreport: An International Journal for the Rapid Communication of Research in Neuroscience, 9,* 1007–1112.

Lippa, Y. (1996). A referential coding explanation of compatibility effects of physically orthogonal stimulus and response dimensions. *Quarterly Journal of Experimental Psychology, 49A,* 950–971.

Logie, R. H. (1995). *Visuo-spatial working memory*. Hove, UK: Lawrence Erlbaum Associates Ltd.

Loomis, J. M., Klatzky, R. L., Golledge, R. G., Cicinelli, J. G., Pellegrino, J. W., & Fry, P. A. (1993). Nonvisual navigation by blind and sighted: Assessment of path integration ability. *Journal of Experimental Psychology: General, 122*, 73–91.

Loomis, J. M., Lippa, Y., Klatzky, R. L., & Golledge, R. G. (2002). Spatial updating of location specified by 3-D sound and spatial language. *Journal of Experimental Psychology: Learning, Memory, and Cognition, 28*, 335–345.

Lund, F. H. (1930). Physical asymmetries and disorientation. *American Journal of Psychology, 42*, 51–62.

MacDonald, J., & McGurk, H. (1978). Visual influence on speech perception processes. *Perception & Psychophysics, 24*, 253–257.

Mack, A., Heuer, F., Villardi, K., & Chambers, D. (1985). The dissociation of position and extent in the Mueller-Lyer figure. *Perception & Psychophysics, 37*, 335–344.

Maculoso, E., Frith, C., & Driver, J. (2000). Modulation of human visual cortex by crossmodal spatial attention. *Science, 289*, 1206–1208.

Maioli, C., & Ohgaki, T. (1993). Role of the accessory optic systems for sensorimotor integration. The problem of reference frames. In A. Berthoz (Ed.), *Multisensory control of movement* (pp. 73–111). Oxford, UK: Oxford University Press.

Malone, P. (1971). Transfer of practice effects with the Müller-Lyer illusion. *Psychonomic Science, 22*, 93–94.

Marchetti, F. M., & Lederman, S. J. (1983). The haptic radial-tangential effect: Two tests of Wong's "moments-of-inertia" hypothesis. *Bulletin of the Psychonomic Society, 21*, 43–46.

Marmor, G. S., & Zaback, I. A. (1976). Mental rotation by the blind: Does mental rotation depend on visual imagery? *Journal of Experimental Psychology: Perception and Performance, 2*, 515–521.

Massaro, D. W., & Anderson, N. H. (1970). A test of a perspective theory of geometrical illusions. *American Journal of Psychology, 83*, 567–575.

Massaro, D. W., & Friedman, D. (1990). Models of integration given multiple sources of information. *Psychological Review, 97*, 225–252.

Maurer, D., Lewis, T. L., Brent, H. P., & Levin, A. V. (1999). Rapid improvement in the acuity of infants after visual input. *Science, 286*, 108–110.

May, M., & Klatzky, R. L. (2000). Path integration while ignoring irrelevant movement. *Journal of Experimental Psychology: Learning, Memory, and Cognition, 26*, 169–186.

McGurk, H., & McDonald, J. (1976). Hearing lips and seeing voices. *Nature, 264*, 746–748.

McGurk, H., Turnure, C., & Creighton, S. J. (1977). Audio-visual coordination in neonates. *Child Development, 48*, 138–143.

Mehta, Z., & Newcombe, F. (1991). A role for the left hemisphere in spatial processing. *Cortex, 27*, 153–167.

Meltzoff, A. N., & Borton, R. W. (1979). Intermodal matching by human neonates. *Nature, 282*, 403–404.

Meredith, M. A., & Stein, B. E. (1996). Spatial determinants of multisensory integration in cat superior colliculus neurons. *Journal of Neurophysiology, 75*, 1843–1857.

Mesulam, M. M. (1998). From sensation to cognition. *Brain, 121*, 1013–1052.
Millar, S. (1971). Visual and haptic cue utilization by preschool children: The recognition of visual and haptic stimuli presented separately and together. *Journal of Experimental Child Psychology, 12*, 88–94.
Millar, S. (1972a). The effects of interpolated tasks on latency and accuracy of intramodal and crossmodal shape recognition by children. *Journal of Experimental Child Psychology, 96*, 170–175.
Millar, S. (1972b). The development of visual and kinaesthetic judgements of distance. *British Journal of Psychology, 63*, 271–282.
Millar, S. (1972c). Effects of instructions to visualise stimuli during delay on visual recognition by preschool children. *Child Development, 43*, 1073–1075.
Millar, S. (1974). Tactile short-term memory by blind and sighted children. *British Journal of Psychology, 65*, 253–263.
Millar, S. (1975a). Effects of input variables on visual and kinaesthetic matching by children within and across modalities. *Journal of Experimental Child Psychology, 19*, 63–78.
Millar, S. (1975b). Effects of tactual and phonological similarity on the recall of Braille letters by blind children. *British Journal of Psychology, 66*, 193–201.
Millar, S. (1975c). Effects of phonological and tactual similarity on serial object recall by blind and sighted children. *Cortex, 11*, 170–180.
Millar, S. (1975d). Spatial memory by blind and sighted children. *British Journal of Psychology, 66*, 449–459.
Millar, S. (1976). Spatial representation by blind and sighted children. *Journal of Experimental Child Psychology, 21*, 460–479.
Millar, S. (1977a). Tactual and name matching by blind children. *British Journal of Psychology, 68*, 377–387.
Millar, S. (1977b). Early stages of tactual matching. *Perception, 6*, 333–343.
Millar, S. (1978). Aspects of information from touch and movement. In G. Gordon (Ed.), *Active touch*. New York: Pergamon Press.
Millar, S. (1979). Utilization of shape and movement cues in simple spatial tasks by blind and sighted children. *Perception, 8*, 11–20.
Millar, S. (1981a). Self-referent and movement cues in coding spatial location by blind and sighted children. *Perception, 10*, 255–264.
Millar, S. (1981b). Crossmodal and intersensory perception and the blind. In R. D. Walk & H. L. Pick, Jr. (Eds.), *Intersensory perception and sensory integration*. New York & London: Plenum Press.
Millar, S. (1984a). Strategy choices by young Braille readers. *Perception, 3*, 567–579.
Millar, S. (1984b). Is there a "best hand" for braille? *Cortex, 20*, 75–87.
Millar, S. (1985a). Movement cues and body orientation in recall of location by blind and sighted children. *Quarterly Journal of Experimental Psychology, 37*, 257–279.
Millar, S. (1985b). The perception of complex patterns by touch. *Perception, 14*, 293–303.
Millar, S. (1986). Aspects of size, shape and texture in touch: Redundancy and interference in children's discrimination of raised dot patterns. *Journal of Child Psychology and Psychiatry, 27*, 367–381.
Millar, S. (1987a). Perceptual and task factors in fluent braille. *Perception, 16*, 521–536.

Millar, S. (1987b). The perceptual "window" in two-handed braille: Do the left and right hands process text simultaneously? *Cortex, 23*, 111–222.

Millar, S. (1988a). Models of sensory deprivation: The nature/nurture dichotomy and spatial representation in the blind. *International Journal of Behavioural Development, 11*, 69–87.

Millar, S. (1988b). An apparatus for recording handmovements. *British Journal of Visual Impairment and Blindness, 6*, 87–90.

Millar, S. (1988c). Prose reading by touch: The role of stimulus quality, orthography and context. *British Journal of Psychology, 79*, 87–103.

Millar, S. (1990). Articulatory coding in prose reading: Evidence from braille on changes with skill. *British Journal of Psychology, 81*, 205–219.

Millar, S. (1994). *Understanding and representing space: Theory and evidence for studies with blind and sighted children*. Oxford, UK: Oxford University Press.

Millar, S. (1997). *Reading by touch*. London: Routledge.

Millar, S. (1999a). Memory in touch. *Psicothema, 11*, 747–767.

Millar, S. (1999b). Veering revisited: Noise and posture cues when walking without sight. *Perception, 28*, 765–780.

Millar, S. (2000). Modality and mind: Convergent active processing in interrelated networks: A model of development and perception by touch. In M. A. Heller (Ed.), *Touch, representation and blindness* (pp. 99–141). Oxford, UK: Oxford University Press.

Millar, S. (2005). Network models for haptic perception. *Infant Behavior and Development, 28*, 250–265.

Millar, S., & Al-Attar, Z. (2000). Vertical and bisection bias in active touch. *Perception, 29*, 481–500.

Millar, S., & Al-Attar, Z. (2001). Illusions in reading maps by touch: Reducing distance errors. *British Journal of Psychology, 92*, 643–657.

Millar, S., & Al-Attar, Z. (2002). Müller-Lyer illusions in touch and vision: Implications for multisensory processes. *Perception & Psychophysics, 64*, 353–365.

Millar, S., & Al-Attar, Z. (2003a). How do people remember spatial information from tactile maps? *British Journal of Visual Impairment, 21*, 64–72.

Millar, S., & Al-Attar, Z. (2003b). Spatial reference and scanning with the left and right hand. *Perception, 32*, 1499–1511.

Millar, S., & Al-Attar, Z. (2004). External and body-centred frames of reference in spatial memory: Evidence from touch. *Perception & Psychophysics, 66*, 51–59.

Millar, S., & Al-Attar, Z. (2005). What aspects of vision facilitate haptic processing? *Brain and Cognition, 59*, 258–268.

Millar, S., & Ittyerah, M. (1992). Movement imagery in young and congenitally blind children: Mental practice without visuo-spatial information. *International Journal of Behavioral Development, 15*, 125–146.

Miller, G. A. (1956). The magical number seven plus or minus two: Some limits on capacity for processing information. *Psychological Review, 63*, 81–87.

Milner, A. D., & Bryant, P. E. (1970). Cross-modal matching by young children. *Journal of Comparative and Physiological Psychology, 71*, 453–458.

Milner, B., & Taylor, L. (1972). Right hemisphere superiority in tactile pattern recognition after cerebral commissurotomy. *Neuropsychologica, 10*, 1–15.

Minami, K., Hay, V., Bryden, M. P., & Free, T. (1994). Laterality effects for tactile patterns. *International Journal of Neuroscience, 74,* 55–69.

Morris, R. G. M. (1989). Does synaptic plasticity play a role in information storage in the vertebrate brain? In R. G. M. Morris (Ed.), *Parallel distributed processing: Implications or psychology and neurobiology.* Oxford, UK: Clarendon Press.

Moses, F. L., & DeSisto, M. J. (1970). Arm-movement responses to Müller-Lyer stimuli. *Perception & Psychophysics, 8,* 376–378.

Mountcastle, V. B. (2005). *The sensory hand: Neural mechanisms of somatic sensations.* Cambridge, MA: Harvard University Press.

Munakata, Y., & McClelland, J. L. (2003). Connectionist models of development. *Developmental Science, 6,* 413–429.

Munsinger, H. (1967). Developing perception and memory for stimulus redundancy. *Journal of Experimental Child Psychology, 5,* 39–49.

Nelson, B., & Greene, E. (1998). Similar effects from operant and comparison response modes. *Perceptual and Motor Skills, 86,* 499–511.

Newcombe, N., & Liben, L. S. (1982). Barrier effects in the cognitive maps of children and adults. *Journal of Experimental Child Psychology, 34,* 46–58.

Newport, R., Rabb, B., & Jackson, S. R. (2002). Noninformative vision improves haptic spatial perception. *Current Biology, 12,* 1661–1664.

Nickerson, R. S. (1965). Response time to same–different judgments. *Perceptual & Motor Skills, 20,* 15–18.

Nicoletti, R., Umiltà, C., & Ladavas, E. (1984). Compatibility due to the coding of relative position of the effectors. *Acta Psychologica, 57,* 133–143.

Nicolls, M. E. R., & Lindell, A. K. (2000). A left hemisphere, but not right hemispace, advantage for tactual simultaneity judgments. *Perception & Psychophysics, 62,* 717–725.

Nolan, C. Y., & Kederis, C. J. (1969). Perceptual factors in braille word recognition. *American Foundation for the Blind Research Series, No. 20.* New York: American Foundation for the Blind.

Norman, J. (2002). Two visual systems and two theories of perception: An attempt to reconcile the constructivist and ecological approaches. *Behavioral & Brain Sciences, 25,* 73–144.

Novikova, L. A. (1973). Blindness and the electrical activity of the brain. *American Foundation for the Blind Research Series, No. 23.* New York: American Foundation for the Blind.

O'Keefe, J. (1991). The hippocampal cognitive map and navigational strategies. In J. Paillard (Ed.), *Brain and space* (pp. 273–295). Oxford, UK: Oxford University Press.

O'Keefe, J., & Nadel, L. (1978). *The hippocampus as a cognitive map.* Oxford, UK: Clarendon Press.

Over, R. (1967). Haptic judgments of the Müller-Lyer illusion by subjects of different ages. *Psychonomic Science, 9,* 365–366.

Over, R. (1968). The effect of instructions on visual and haptic judgment of the Müller-Lyer illusion. *Australian Journal of Psychology, 20,* 161–164.

Paillard, J. (1991). Motor and representational framing of space. In J. Paillard (Ed.), *Brain and space* (pp. 163–184). Oxford, UK: Oxford University Press.

Pascual-Leone, A., Theoret, H., Merabet, L., Kauffmann, T., & Schlaug, G. (2006). The role of visual cortex in tactile processing: A metamodal brain. In M. A. Heller & S. Ballesteros (Eds.), *Touch and blindness: Psychology and neuroscience* (pp. 171–195). Mahwah, NJ: Lawrence Erlbaum Associates Inc.

Pascual-Leone, A., & Torres, F. (1993). Plasticity of the sensorimotor cortex representation of the reading finger in Braille readers. *Brain, 116,* 39–52.

Pascual-Leone, A., Wassermann, E., Sadato, N., & Hallett, M. (1995). The role of reading activity on the modulation of motor outputs to the reading hand in braille readers. *Annals of Neurology, 38,* 910–915.

Pashler, H. (1990). Coordinate frame for symmetry detection and object recognition. *Journal of Experimental Psychology: Human Perception and Performance, 16,* 150–163.

Patterson, J., & Deffenbacher, K. (1972). Haptic perception of the Müller-Lyer illusion by the blind. *Perception & Motor Skills, 35,* 819–824.

Pavlov, I. P. (1927). *Conditioned responses.* New York: Oxford University Press.

Pearce, H.-J. (1904). The law of attraction in relation to some visual and tactual illusions. *Psychological Review, 11,* 143–178.

Pepper, R., & Herman, L. M. (1970). Decay and interference in the short-term retention of a discrete motor act. *Journal of Experimental Psychology, Monograph Series, 83,* no. 2.

Peterson, L. R., & Peterson, M. J. (1959). Short-term retention of individual verbal items. *Journal of Experimental Psychology, 58,* 193–198.

Piaget, J. (1953). *Logic and psychology.* Manchester, UK: Manchester University Press.

Piaget, J. (1954). *The construction of reality in the child.* New York: Basic Books.

Piaget, J. (1969). *The mechanisms of perception* (G. N. Seagrin, Trans.). London: Routledge & Kegan Paul.

Piaget, J., & Inhelder, B. (1956). *The child's conception of space.* London: Routledge & Kegan Paul.

Pietrini, P., Furey, M. L., Ricciardi, E., Gobbini, M. I., Wu, W. H. C., Cohen, L., et al. (2004). Beyond sensory images: Object-based representations in the human ventral pathway. *Proceedings of the National Academy of Sciences of the United States of America, 101,* 5659–5663.

Pochon, J. P., Levy, R., Poline, J. B., Crozier, S., Lehheriscy, S., Pillon, B., et al. (2001). The role of dorsolateral prefrontal cortex in the preparation of forthcoming action: An fMRI study. *Cerebral Cortex, 11,* 260–266.

Pont, S. C., Kappers, A. M. L., & Koenderink, J. J. (1999). Similar mechanisms underlie curvature comparison by static and by dynamic touch. *Perception & Psychophysics, 61,* 874–894.

Pouget, A., & Snyder, L. H. (2000). Computational approaches to sensorimotor transformations. *Nature Neuroscience, 3,* 1192–1106.

Predebon, J. (1992). Framing effects and the reversed Müller-Lyer illusion. *Perception and Psychophysics, 52,* 307–314.

Predebon, J. (1996). The role of the angle components in the wings-in and wings-out forms of the Müller-Lyer illusion. *Perception, 25,* 773–781.

Predebon, J. (1998). Decrement of the Brentano Müller-Lyer illusion as a function of inspection time. *Perception, 27,* 183–192.

Pressey, A. (1974). Effects of size of angle on the ambiguous Müller-Lyer illusion. *Acta Psychologica, 38,* 401–404.
Pressey, A. W., & Pressey, C. A. (1992). Attentive fields are related to focal and contextual features: A study of Müller-Lyer distortions. *Perception & Psychophysics, 51,* 423–436.
Prinzmetal, W., & Gettleman, L. (1993). Vertical–horizontal illusion: One eye is better than two. *Perception & Psychophysics, 53,* 81–88.
Proctor, R. W., & Reeve, T. G. (Eds.). (1990). *Stimulus–response compatibility: An integrated perspective.* Amsterdam: North-Holland.
Rains, G. D., & Milner, B. (1994). Right-hippocampal contralateral hand effect in the recall of spatial location in the tactual modality. *Neuropsychologia, 32,* 1233–1242.
Rauschecker, J. P., Tian, B., Korte, M., & Egert, U. (1992). Crossmodal changes in the somatosensory vibrissae/barrel system of visually deprived animals. *Proceedings of the National Academy of Sciences of the United States of America, 89,* 5063–5069.
Recanzone, G. H., Merzenich, M. M., Jenkins, W. M., Grajski, K. A., & Dinse, H. R. (1992). Topographic reorganization of the hand representation in cortical area 3b of owl monkeys trained in a frequency discrimination task. *Neurophysiology, 67,* 1031–1056.
Reid, R. L. (1954). An illusion of movement complementary to the horizontal–vertical illusion. *Quarterly Journal of Experimental Psychology, 6,* 107–111.
Revesz, G. (1950). *Psychology and art of the blind.* London: Longmans.
Révész, G. (1934). System der optischen und haptischen Raumtäuschungen. *Zeitschrift für Psychologie, 131,* 125–131.
Richardson, B., & Wuillemin, D. B. (1981). Different orientations of sub-two-point threshold tactile stimuli can be discriminated. *Bulletin of the Psychonomic Society, 18,* 311–314.
Rieber, C. H. (1903). Tactual illusions. *Psychological Monographs, 4,* 47–99.
Riesen, A. H. (1947). The development of visual perception in man and chimpanzee. *Science, 106,* 107–108.
Rieser, J. (1979). Reference systems and the spatial orientation in six-month-old infants. *Child Development, 50,* 1078.
Rieser, J., Ashmead, C. R. T., & Youngquist, G. A. (1990). Visual perception and the guidance of locomotion without vision to previously seen targets. *Perception, 19,* 675–689.
Rieser, J., Guth, D. A., & Hill, E. W. (1986). Sensitivity to perspective structure while walking without vision. *Perception, 15,* 137–188.
Rieser, J., & Heiman, M. L. (1982). Spatial self-reference systems and shortest route behavior in toddlers. *Child Development, 53,* 524–533.
Rieser, J., Lockman, J. J., & Pick, H.-L. Jr. (1980). The role of visual experience in knowledge of spatial lay-out. *Perception & Psychophysics, 28,* 185–190.
Rieser, J. J., & Rider, E. A. (1991). Young children's spatial orientation with respect to multiple targets when walking without vision. *Developmental Psychology, 27,* 97–107.
Riggio, L., Dellantonio, A., & Barbato, M. (1980). Right hand superiority in serial recall of tactile stimuli. *Cortex, 18,* 595–601.

Rizzolatti, G., Fogassi, L., & Gallese, V. (1997). Parietal cortex from sight to action. *Current Opinion in Neurobiology, 7*, 562–567.

Rizzolatti, G., Fogassi, L., & Gallese, V. (2004). Cortical mechanisms subserving object grasping, action understanding and imitation. In M. S. Gazzaniga (Ed.), *The cognitive neurosciences* (3rd ed., pp. 427–440). Cambridge, MA: MIT Press.

Robertson, A. (1902). Studies from the psychological laboratory of the University of California VI: Geometric-optical illusions in touch. *Psychological Review, 9*, 549–569.

Rock, I. (1973). *Orientation and form*. London: Academic Press.

Rock, I. (1984). *Perception*. New York: W. H. Freeman.

Rock, I. (1997). *Indirect perception*. Cambridge, MA: MIT Press.

Rock, I., & Victor, J. (1963). Vision and touch: An experimentally created conflict between two senses. *Science, 143*, 594–596.

Röder, B., Kusmierek, A., Spence, C., & Schicke, T. (2007). Developmental vision determines the reference frame for multisensory control of action. *Proceedings of the National Academy of Sciences of the United States of America. 104*, 4753–4758.

Röder, B., Teder-Sälejärvi, W., Sterr, A., Rösler, F., Hillyard, S. A., & Neville, H. (1999). Improved auditory spatial tuning in blind humans. *Nature, 400*, 162–166.

Rolls, E. T. (1991). Functions of the primate hippocampus in spatial processing and memory. In J. Paillard (Ed.), *Brain and space* (pp. 353–378). Oxford, UK: Oxford University Press.

Rorden, C., Heutink, J., Greenfield, E., & Robertson, I. H. (1999). When a rubber hand "feels" what the real hand cannot. *NeuroReport, 10*, 135–138.

Rösler, F., Röder, B., Heil, M., & Henninghausen, F. (1993). Topographical differences of slow, event-related brain potentials in blind and sighted adult human subjects during haptic mental rotation. *Brain Research: Cognitive Brain Research, 1*, 145–159.

Ross, H. (1974). *Behaviour and perception in strange environments. Advances in Psychology Series 5*. London: George Allen & Unwin.

Rouse, D. L., & Worchel, P. (1955). Veering tendency in the blind. *New Outlook for the Blind, 49*, 115–118.

Rudel, R. G., Denckla, M. B., & Hirsch, S. (1977). The development of left hand superiority for discriminating Braille configurations. *Neurology, 27*, 160–164.

Rudel, R. G., & Teuber, H.-L. (1963). Decrement of visual and haptic Müller-Lyer illusion on repeated trials: A study of crossmodal transfer. *Quarterly Journal of Experimental Psychology, 15*, 125–131.

Rudel, R. G., & Teuber, H.-L. (1964). Crossmodal transfer of shape discrimination by children. *Neuropsychologia, 2*, 1–8.

Rumelhart, D. E., & McClelland, J. L. (1986). *Parallel distributed processing: Explorations in the microstructure of cognition* (Vols. 1 & 2). Cambridge, MA: MIT Press.

Rushworth, M. F. S., Ellison, A., & Walsh, V. (2001). Complementary localisation and lateralization of orienting and motor attention. *Nature Neuroscience, 4*, 656–661.

Rushworth, M. F. S., Nixon, P. D., & Passingham, R. E. (1997a). Parietal cortex and movement: I. Movement selection and reaching. *Experimental Brain Research, 117*, 292–310.

Rushworth, M. F. S., Nixon, P. D., & Passingham, R. E. (1997b). Parietal cortex and movement: II. Spatial representation. *Experimental Brain Research, 117*, 311–323.

Rushworth, M. F. S., Nixon, P. D., Renowden, S., Wade, D. T., & Passingham, R. E. (1997). The left parietal cortex and motor attention. *Neuropsycholgia, 35*, 1261–1273.

Sabate, M., Gonzales, B., & Roderiguez, M. (2004). Brain lateralization of motor imagery: Motor planning asymmetry as a cause of movement lateralization. *Neuropsychologia, 42*, 1041–1049.

Sadato, N., Pascual-Leone, A., Grafman, J., Deiber, M. P., Ibanez, V., & Hallett, M. (1998). Neural networks for Braille reading by the blind. *Brain, 121*, 1213–1229.

Sadato, N., Pascual-Leone, A., Grafman, J., Ibanez, V., Deiber, M. P., Dold, G., et al. (1996). Activation of the primary visual cortex in braille reading by blind subjects. *Nature, 380*, 526–528.

Sakata, H., & Iwamura, Y. (1978). Cortical processing in the first somatosensory and parietal association areas in the monkey. In G. Gordon (Ed.), *Active touch: The mechanism of recognition of objects by manipulation. A multidisciplinary approach* (pp. 55–72). Oxford, UK: Pergamon Press.

Sathian, K. (2000). Practice makes perfect: Sharper tactile perception in the blind. *Neurology, 54*, 2203–2206.

Sathian, K., & Zangaladze, A. (1996). Tactile spatial acuity at the human fingertip and lip: Bilateral symmetry and inter-digit variability. *Neurology, 46*, 1464–1466.

Schaeffer, A. A. (1928). Spiral movement in man. *Journal of Morphology, 45*, 293–298.

Schlaegel, T. F. (1953). The dominant method of imagery in blind compared to sighted adolescents. *Journal of Genetic Psychology, 83*, 265–277.

Schluter, N. D., Krams, M., Rushworth, M. F. S., & Passingham, R. E. (2001). Cerebral dominance for action in the human brain. *Neuropsychologia, 39*, 105–113.

Semenza, C., Zoppello, M., Gidiuli, O., & Borgo, F. (1996). Dichaptic scanning of Braille letters by skilled blind readers. *Perceptual and Motor Skills, 82*, 1071–1074.

Senden, M. von (1932). *Raum und Gestalt: Auffassung bei operierten Blindgeborenen vor und nach der Operation.* Leipzig: Barth.

Serrati, C., Finocchi, C., Calautti, C., Bruzzone, G. L., Colucci, M., Gandolfo, C., et al. (2000). Absence of hemispheric dominance for mental rotation ability: A transcranial Doppler study. *Cortex, 36*, 415–425.

Shatz, C. J. (1992). The developing brain. *Scientific American, 267*, 61–67.

Shepard, R. N., & Cooper, L. A. (1982). *Mental images and their transformations.* Cambridge, MA: MIT Press.

Shepard, R. N., & Feng, C. (1972). A chronometric study of mental paperfolding. *Cognitive Psychology, 3*, 228–243.

Shepard, R. N., & Metzler, J. (1971). Mental rotation of three-dimensional objects. *Science, 171*, 701–703.

Shiffrin, R. M., & Schneider, W. (1977). Controlled and automatic human information processing: II. Perceptual learning, automatic attending and a general theory. *Psychological Review, 84*, 127–190.

Siegel, A., & White, S. H. (1975). The development of spatial representation of large-scale environments. In H. W. Reese (Ed.), *Advances in child development & behavior* (Vol. 10). New York: Academic Press.

Simon, J. R., & Rudell, A. P. (1967). Auditory S-R compatibility: The effect of an irrelevant cue on information processing. *Journal of Applied Psychology, 51*, 300–304.

Slator, R. (1982). *The development of spatial perception and understanding in young children*. D.Phil. Thesis, University of Oxford, UK.

Smith, M. O., Chu, J., & Edmonston, W. E. Jr. (1977). Cerebral lateralization of haptic perception: Interaction of responses to braille and music reveals a functional basis. *Science, 197*, 689–690.

Smyth, M. M., Hay, D. C., Hitch, G. J., & Horton, N. J. (2005). Serial position memory in the visuo-spatial domain: Reconstruction sequences of unfamiliar faces. *Quarterly Journal of Experimental Psychology, 58A*, 909–930.

Smyth, M. M., & Pendleton, L. R. (1990). Space and movement in working memory. *Quarterly Journal of Experimental Psychology, 42A*, 291–304.

Smyth, M. M., & Scholey, K. A. (1996). Serial order in spatial immediate memory. *Quarterly Journal of Experimental Pychology, 49A*, 159–177.

Snyder, L. H., Grieve, K. L., Brotchie, P., & Andersen, R. A. (1998). Separate body and world-referenced representations of visual space in the parietal cortex. *Nature, 394*, 887–891.

Spence, C., & Driver, J. (Eds.). (2004). *Crossmodal space and crossmodal attention*. Oxford, UK: Oxford University Press.

Spivey, K. M. J., & Bridgeman, B. (1993). Spatial context affects the Poggendorff illusion. *Perception & Psychophysics, 53*, 467–474.

Springbett, B. M. (1961). Some stereoscopic phenomena and their implications. *British Journal of Psychology, 25*, 105–109.

Stein, B. E., & Meredith, M. A. (1993). *The merging of the senses*. Cambridge, MA: MIT Press.

Stein, B. E., Wallace, M. T., & Stanford, T. R. (2000). Merging sensory signals in the brain: The development of multisensory integration in the superior colliculus. In M. S. Gazzaniga (Ed.), *The new cognitive neuroscience* (pp. 55–71). Cambridge, MA: MIT Press.

Stein, J. F. (1991). Space and the parietal association areas. In J. Paillard (Ed.), *Brain and space* (pp. 185–222). Oxford, UK: Oxford University Press.

Stein, J. F. (1992). The representation of egocentric space in the posterior parietal cortex. *Behavioral and Brain Sciences, 15*, 691–700.

Stevens, J. C., Foulke, E., & Patterson, M. Q. (1996). Tactile acuity, aging, and Braille reading in long-term blindness. *Journal of Applied Psychology, 2*, 91–106.

Steven, M. S., & Blakemore, C. (2004). Cortical plasticity in the adult human brain. In M. Gazzaniga (Ed.), *The cognitive neurosciences* (pp. 1243–1254). Cambridge. MA: MIT Press.

Strelow, E. R., & Brabyn, J. A. (1982). Locomotion of the blind controlled by natural sound cues. *Perception, 11*, 635–640.

Strelow, E. R., Brabyn, J. A., & Clark, G. R. S. (1976). Apparatus for measuring and recording path velocity and direction characteristics of human locomotion. *Behavior Research Methods & Instrumentation, 8*, 442–446.

Streri, A. (1987). Tactile perception of shape and intermodal transfer in 2- to three-month-old infants. *British Journal of Developmental Psychology, 5,* 213–220.
Streri, A. (1993). *Seeing, reaching, touching: The relation between vision and touch in infancy.* Cambridge, MA: MIT Press.
Streri, A. (2000). Intermodal relations in infancy. In Y. Hatwell, A. Streri, & E. Gentaz (Eds.), *Touching for knowing: Cognitive psychology of haptic manual perception* (pp. 191–206). Amsterdam/Philadelphia: John Benjamins.
Streri, A., & Gentaz, E. (2003). Cross-modal recognition from hand to eyes in human newborns. *Somatosensory & Motor Research, 20,* 11–16.
Suzuki, K., & Arashida, R. (1992). Geometric haptic illusions revisited: Haptic illusions compared with visual illusions. *Perception & Psychophysics, 52,* 329–335.
Tinti, C., Adenzato, M., Tamietto, M., & Cornoldi, C. (2006). Visual experience is not necessary for efficient survey spatial cognition: Evidence from blindness. *Quarterly Journal of Experimental Psychology, 59,* 1306–1328.
Tipper, S. P. (2004). Attention and action. In M. S. Gazzaniga (Ed.), *The cognitive neurosciences* (pp. 619–629). Cambridge, MA: MIT Press.
Tipper, S. P., Lloyd. D., Shorland, B., Howard, L. A., & McGlone, F. (1998). Vision influences tactile perception without proprioceptive orienting. *NeuroReport, 9,* 1741–1744.
Titchener, E. B. (1909). *Lectures on the experimental psychology of thought.* New York: MacMillan.
Tobin, M. J. (1988). *Beginning Braille: Self instruction course* (Revised ed.). London: Royal National Institute for the Blind.
Tolman, E. C. (1948). Cognitive maps in rats and men. *Psychological Review, 40,* 60–70.
Treisman, A. (2004). Psychological issues in selective attention. In M. S. Gazzaniga (Ed.), *The cognitive neurosciences.* Cambridge, MA: MIT Press.
Tsai, L. S. (1967). Müller-Lyer illusion by the blind. *Perceptual and Motor Skills, 25,* 641–644.
Turnbull, O. H., Driver, J., & McCarthy, R. A. (2004). 2D but not 3D: Pictorial-depth deficits in a case of visual agnosia. *Cortex, 40,* 723–737.
Tversky, B. (1981). Distortion in memory for maps. *Cognitive Psychology, 13,* 407–433.
Tversky, B., & Schiano, D. J. (1989). Perceptual and conceptual factors in distortions in memory for graphs and maps. *Journal of Experimental Psychology: General, 118,* 387–398.
Ungeleider, L. G., & Mishkin, M. (1982). Two cortical visual systems. In D. Ingle, A. Goodale, & R. J. W. Mansfield (Eds.), *The analysis of visual behaviour* (pp. 549–586). Cambridge, MA: MIT Press.
Vallar, G., Lobel, E., Galati, G., Berthoz, A., Pizzamiglio, L., & Le-Bihan, D. (1999). A fronto-parietal system for computing the egocentric spatial frame of reference in humans. *Experimental Brain Research, 124,* 281–286.
Veraart, C., DeVolder, A., Wanet-Defalque, M., Bol, A., Michel, C., & Goffinet, A. (1990). Glucose utilisation in human visual cortex is abnormally elevated in blindness of early onset but decreased in blindness of late onset. *Brain Research, 510,* 115–121.

Vurpillot, E. (1976). *The visual world of the child*. London: George Allen & Unwin.
Walker, J. T. (1971). Visual capture in visual illusions. *Perception & Psychophysics, 10*, 71–74.
Wallace, M. T., & Stein, B. E. (2007). Early experience determines how the senses will interact. *Journal of Neurophysiology, 97*, 921–926.
Wallace, R. J. (1972). Spatial S-R compatibility effects involving kinesthetic cues. *Journal of Experimental Psychology, 93*, 163–168.
Wallace, S. A. (1977). The coding of location: A test of the target hypothesis. *Journal of Motor Behavior, 9*, 343–357.
Wanet-Defalque, M., Veraart, C., DeVolder, A., Metz, R., Dooms, G., et al. (1988). High metabolic activity in the visual cortex of early blind human subjects. *Brain Research, 446*, 369–373.
Warren, D. H. (1977). *Blindness and early childhood development*. New York: American Foundation for the Blind.
Watt, H. J. (1905). Experimentelle Beiträge zu einer Theorie des Denkens. *Archiv der gesammelten Psychologie, 4*, 285–436.
Weber, E. H. (1834 & 1846/1978). *De Tactu* [The sense of touch] (H. E. Ross, Trans.) & *Der Tastsinn und das Gemeingefüh* [The sense of touch and common sensibility] (D. J. Murray, Trans.). In: *The sense of touch*. London: Academic Press, 1978.
Weeks, D. J., & Proctor, R. W. (1990). Salient features coding in the translation between orthogonal stimulus–response dimensions. *Journal of Experimental Psychology: General, 119*, 355–366.
Weeks, D. J., Proctor, R. W., & Beyak, B. (1995). Stimulus–response compatibility for vertically oriented stimuli and horizontally oriented responses. *Quarterly Journal of Experimental Psychology, 48A*, 367–383.
Weiskrantz, L. (1986). *Blindsight: A case study and its implications*. Oxford, UK: Oxford Science Publications.
Welsh, R. A., & Blash, B. B. (1980). *Foundations of orientation and mobility*. New York: American Foundation for the Blind.
Wenderoth, P. (1992). Perceptual illusions. *Australian Journal of Psychology, 44*, 147–151.
Wenderoth, P., & Alais, D. (1990). Lack of evidence for a tactual Poggendorff illusion. *Perception & Psychophysics, 48*, 234–242.
Wenderoth, P., & Wade, N. (1981). An investigation of line and dot forms of the Müller-Lyer and Poggendorff illusions. *Quarterly Journal of Experimental Psychology, 33A*, 77–85.
Wertheimer, M. (1912/1932). Experimentelle Studien über das Sehen von Bewegung. *Zeitschrift für Psychologie, 61*, 161–265.
Wertheimer, M. (1961). Psychomotor coordination of auditory input ands visual space at birth. *Science, 134*, 1696.
Williams, P. A., & Enns, J. T. (1996). Pictorial depths and framing have independent effects on the horizontal–vertical illusion. *Perception, 25*, 921–926.
Wolfe, U., Maloney, T., & Tam, M. (2005). Distortions of perceived length in the frontoparallel plane: Tests of perspective theories. *Perception & Psychophysics, 67*, 967–979.

Wong, T. S. (1975). The respective role of limb and eye movements in the haptic and visual Müller-Lyer illusion. *Quarterly Journal of Experimental Psychology, 27*, 659–666.

Wong, T. S. (1977). Dynamic properties of radial and tangential movements as determiners of the horizontal–vertical illusion. *Journal of Experimental Psychology: Human Perception and Performance, 3*, 151–164.

Wong, T. S., Ho, R., & Ho, J. (1974). Influence of shape of receptor organ on the horizontal vertical illusion. *Journal of Experimental Psychology, 103*, 414–419.

Woodin, M. E., & Allport, A. (1998). Independent reference frames in human spatial memory: Body-centred and environment-centered coding in near and far space. *Memory & Cognition, 26*, 1109–1116.

Worchel, P. (1951). Space perception and orientation in the blind. *Psychological Monographs, 65*, Whole no. 332.

Wundt, W. (1862). *Beiträge zur Theorie der Sinneswahrnehmung* [Contributions to the theory of sensory perception]. Leipzig & Heidelberg: C. F. Winter.

Wundt, W. (1898). Die geometrisch-optischen Täuschungen [The geometric-optical illusions]. *Abhandlungen der Mathematisch-Physischen Klasse der Königlich-Sächsischen Gesellschaft der Wisssenschaften, 24*, 53–178.

Zangaladze, A., Epstein, C. M., Grafton, S. T., & Sathian, K. (1999). Involvement of visual cortex in tactile discrimination of orientation. *Nature, 401*, 587–590.

Zhou, R., & Black, I. B. (2000). The development of neural maps: Molecular mechanisms. In M. S. Gazzaniga (Ed.), *The new cognitive neurosciences* (2nd ed., pp. 213–236). Cambridge, MA: MIT Press.

Zhou, Y. D., & Fuster, J. M. (2004). Somatosensory cell response to an auditory cue in a haptic memory task. *Behavioural Brain Research, 153*, 573–578.

Author index

Abrams, R. A. 116
Adam, J. J. 44
Adams, J. A. 8, 102
Adenzato, M. 36
Aggleton, J. P. 44, 106
Alais, D. 159
Al-Attar, Z. 90, 91, 96, 108, 126, 134, 138, 139, 144, 149, 154, 155, 172, 173, 190
Alho, K. 41, 184
Ahonen, A. I. 41
Akbudak, E. 81
Allport, A. 104
Allyn, M. R. 148
Amedi, A. 81, 169, 179, 184
Andersen, R. A. 32, 33, 105, 165, 169
Anderson, N. H. 146
Andrew, H. 101, 125
Andrew, R. 50, 179
Anii, A. 149, 156
Anker, S. 50, 179
Appelle, S. 72, 102
Arashida, R. 159
Arbib, M. 34, 105
Arieh, Y. 123
Armstrong, L. 123, 126
Ashmead, C. R. T. 48

Ashmead, D. H. 49
Atkinson, J. 50, 179
Atkinson, R. C. 21
Avery, G. C. 123

Baddeley, A. D. 22, 25, 29, 30, 73, 116
Ballesteros, S. 30, 76, 116, 138
Banks, M. S. 31, 164
Barber, P. O. 102
Barbato, M. 89
Bardisa, D. 30, 116
Barnette, S. L. 85
Barrett, D. J. K. 101
Bauer, J. A. 15, 28
Bayliss, G. C. 34, 105, 106
Bean, C. H. 119, 145
Beers, R. J. van 31, 37, 163, 164
Behrmann, M. 87, 90
Berlá, E. P. 89
Bertelson, P. 82
Berthoz, A. 33, 44, 87, 90, 101, 105, 106
Beyak, B. 162
Bi, G.-G. 22, 23
Bisiach, E. 87, 89
Black, I. B. 181
Blakemore, C. 184
Blanco, F. 145

Blash, B. B. 67
Blundell, J. 101, 125
Bodis-Wollner, I. 8
Bol, A. 41
Bolles, R. C. 148.
Borgo, F. 88
Borton, R. W. 28
Bottini, G. 165,
Boven, R. W. van 82, 185
Boyer, A. 145
Boynton, G. M. 146
Brabyn, J. A. 49, 55
Brackett, D. D. 120, 122, 145, 146
Braddick, O. 50, 179
Bradley, D. C. 32
Bradshaw, J. L. 87
Bradshaw, M. F. 101
Brainard, G. C. 179
Brent, H. P. 35, 41
Brewer, A. A. 146
Bridgeman, B. 154
Broadbent, D. E. 21
Brooks, J. 69
Brooks, L. R. 21
Brotchie, P. 105, 165
Brown, J. A. 21
Bruzzone, G. L. 89
Bryant, P. E. 28, 29
Bryden, M. P. 89
Büchel, C. 81, 169
Buelthoff, H. H. 31
Bürklen, K. 73, 80, 82
Burnod, Y. 34
Burson, L. L. 124, 135
Burton, H. 81
Burtt, H. E. 119
Butterfield, L. H. Jr. 89, 90
Butterworth, G. 28

Calautti, C. 89
Calcaterra, J. A. 85, 124, 135
Calvert, G. 30, 163
Campbell, R. 30
Carlson, S. 41
Carpenter, P. A. 36
Carrasco, M. 147
Cashdan, S. 29
Casla, M. 145
Castillo, M. 28
Catala, M. D. 169

Celnik, P. 168, 169
Cezayirli, C. L. 44, 106
Chambers, D. 149
Chapman, E. K. 96
Chataway, C. 37
Cheng, M. F. H. 124
Cicinelli, J. G. 36, 67
Chieffi, S. 144
Cho, Y. S. 163
Chow, K. L. 9, 41
Chu, J. 87, 89
Claparède, E. 51
Clark, G. R. S. 55, 66
Claxton, V. 28
Cohen, L. 168, 169, 179
Cohen, L. G. 184
Colby, C. L. 32
Collani, G. von 122
Colucci, M. 89
Committeri, G. 106
Connolly, K. 29
Conrad, R. 21
Conturo, T. E. 81
Cooke, F. D. 167
Cooper, L. A. 24
Coren, S. 146, 149, 156
Cornoldi, C. 36
Corwell, B. 169
Countryman, M. 102
Craig, J. C. 167
Cratty, B. J. 51, 52, 67
Creighton, S. J. 28
Cross, J. 29.
Crozier, S. 31, 33
Cuijpers, R. H. 102
Curran, W. 50, 179

D'Alimonte, G. 82.
Dambrosia, J. 168, 169
Dan-Fodio. H. 126, 128
Danziger, S. 44
Daprati, E. 144
Darwin, C. 8
Davenport, R. K. 28
Davidon, R. S. 124
Davidson, P. W. 72, 73
Davis, I. 145
Day, R. H. 123, 124, 147
Deffenbacher, K. 145
Deiber, M. P. 81, 169
Delantionio, A. 89

Della Sala, S. 116
DeLucia, P. 144
Denckla, M. B. 87
Deregowski, J. 124
De Renzi, E. 88
DeSisto, M. J. 149
DeVolder, A. 41
Dewar, R. E. 149
Dijkstra, S. 102
Dinse, H. R. 101
Dodd, B. 30
Dold, G. 81, 169
Dooms, G. 41
Driver, J. 30, 146, 167
Dufosse, M. 34
Duhamel, J.-R. 32
Dunford, M. 87

Eaton, S. B. 49
Ebert, P. C. 143, 147
Ebinger, K. A. 49
Edmonston, W. E. Jr. 87, 89
Egert, U. 184
Eisenberg, P. 36
Ellis, H. D. 124
Ellison, A. 88
Enns, J. T. 122
Enoch, J. M. 8
Epstein, C. M. 184
Erlebacher, A. 148
Ernst, M. O. 31, 164
Ettlinger, G. 29
Evans, G. 36.
Everatt, P. J. 101

Faiz, L. 168, 169
Faglioni, P. 88
Farrand, P. 113
Farrell, M. J. 48
Fellows, B. J. 149, 154
Feng, C. 24
Fertsch, P. 87
Fessard, A. 34.
Festinger, L. 148
Fick, A. 121
Figueroa, J. G. 147
Fine, I. 146
Finger, F. W. 121
Finocchi, C. 89
Fischer, M. H. 44
Fisher, G. H. 146

Fitts, P. M. 162
Fodor, J. 191
Fogassi, L. 34, 167
Foulke, E. 73, 74, 81, 82
Frackowiak, R. S. J. 81, 165, 169
Fraisse, P. 149
Free, T. 89
Friedman, D. 163
Friedman, H. R. 30
Frith, C. D. 165, 167
Frisby, J. 145
Frith, C. 167
Friston, K. 81, 169
Fry, C. L. 145
Fry, P. A. 36
Furedi, L. 40
Furey, M. L. 169, 179
Fuster, J. M. 185

Galati, G. 105, 106
Gallese, V. 34, 167
Gandolfo, C. 89
Gati, J. S. 169
Gazzaniga, M. S. 34, 87
Gentaz, E. 28, 39, 76, 102, 120
Gentilucci, M. 144
Gescheider, G. A. 126
Gettleman, L. 122, 123
Gibson, E. J. 17, 29, 147
Gibson, J. J. 13, 16–19, 29, 69, 73, 88, 147
Gidiuli, O. 88
Gillan, D. J. 139, 154
Girgus, J. 149
Glennerster, A. 146
Gobbini, M. I. 169, 179
Goble, A. K. 149
Goffinet, A. 41
Goldberg, M. E. 32
Goldenberg, G. 37
Goldman-Rakic, P. S. 30
Goldreich, D. 82, 185
Golledge, R. G. 36, 67, 69
Gollin, E. S. 40
Gon, J. J. 37, 163, 164
Gonzales, B. 88
Goodale, M. A. 169
Goodman, D. 146
Goodnow, J. J. 29
Goryo, K. 149, 156
Grafman, J. 81, 169
Grafton, S. T. 184

AUTHOR INDEX

Grajski, K. A. 101
Grant, A. C. 74, 82
Gravetter, F. G. 102
Gray, C. 116
Graziano, M. S. 33, 105, 167
Green, D. M. 23, 167
Green, S. L. 85, 124, 135
Greene, E. 162
Greenfield, E. 28
Gregory, R. L. 35, 122, 145, 146, 148
Grieve, K. L. 105, 165
Griggs, R. 146
Gross, C. G. 33. 105, 167
Grunewald, A. P. 79
Guest, S. 31
Gugerty, L. 69
Guldberg, G. A. 51.
Gurfinkel, V. S. 32 33
Guth, D. 48, 49, 51, 52, 62

Haber, R. N. 72, 73
Haggard, P. 31, 101, 125, 164, 166, 167
Hallett, M. 81
Hamalaien, M. S. 41
Hamilton, R. H. 82, 185
Hanifin, J. P. 179.
Hanowski, R. J. 139, 154
Harris, J. C. 67
Harvey, L. O. Jr. 117.
Hatwell, Y. 39, 76, 102, 120, 123, 124, 131
Hay, D. C. 113
Hay, V. 89
Hebb, D. O. 21, 22
Heil, M. 41
Heiman, M. L. 39
Hein, A. 15
Held, R. 15, 28
Henninghausen, F. 41
Heller, M. A. 31, 73, 85, 88, 120, 122, 124, 126, 128, 135, 145, 146,
Helmholtz, H. von 10, 44, 119, 120, 143
Hendler, T. 81, 169, 179, 184
Henry, L. 25
Herman, L. M. 134
Hermelin, B. 87
Heuer, F. 149
Heutink, J. 27
Hill, E. W. 48
Hillyard, S. A. 41, 185
Hirsch, S. 87

Hitch, G. J. 22, 25, 113
Ho, J. 131
Ho, R. 131
Holdstock, J. S. 44, 106
Hollins, M. 9, 36, 41, 49, 149
Horton, N. J. 113
Howard, L. A. 167
Howard, I. P. 14–16, 28, 44, 52, 69, 121
Hummel, F. C. 184
Humphrey, G. K. 169
Huotilainen, M.. 41, 184
Huttenlocher, P. R. 28
Hyvärinen, J. 41
Hyvärinen, L. 41

Ibanez, V. 81, 169
Igel, A. 117
Ilmoniemi, R. J. 184
Inhelder, B. 148
Isaak, J. P. 44, 106
Ittyerah, M. 30, 102, 116
Iwamura, Y. 32, 33

Jackson, S. R. 163, 164, 166
Jacobson, G. 169, 179, 184
James, T. W. 169
Jastrow, J. 27
Jeannerod, M. 8, 90
Jenkins, W. M. 101
Johnson, K.O. 167
Jolles, J. 44
Jones, B. 29
Jones, D. 113
Jones, P. 28
Jouen, F. 28
Joyner, T. D. 124, 126, 128
Judd, C. H. 145, 146, 147, 148

Kanics I. M. 82, 185
Kant, I. 8, 10
Kappers, A. M. L. 101, 102
Katz, D. 1, 9, 11–13, 18, 72, 88, 100
Katz, L. C. 28, 29
Karnath, H. O. 165
Kauffman, T. 82, 184, 185
Kederis, C. J. 73, 75
Kelley, E. K. 49
Kennett, S. 166, 167
Kennish, J. 144
Keulen, R. F. 44

Kimura, D. 87, 88
Kingstone, A. 44
Klatzky, R. L. 31, 36, 37, 52, 67, 69, 102
Klauer, K. C. 36
Kobor, I. 40
Koenderink, J. J. 101, 102
Koffka, K. 11
Korte, M. 184
Kosslyn, S. M. 24, 37, 39
Kovacs, G. 40
Krams, M. 88
Krauthammer, G. 28
Kress, G. 29
Krueger, L. E. 75
Künnapas, T. M. 121–122, 123, 130, 132
Kudo, K. 149, 156
Kuipers, H. 44
Kujala, T. 41, 184
Kumar, S. 87
Kusajima, T. 80, 82
Kusmierek, A. 40
Kumar, S. 87

Laabs, G. J. 90, 102
Ladavas, E. 162, 163
LaDuke, R. 51, 52, 62
Langdon, D. 89
Lassonde, M. 185
Lawrence, B. M. 116
Le-Bihan, D. 105, 106
LeDoux, J. E. 87
Lechelt, E. C. 88, 102
Lederman, S. J. 31, 37, 102, 124
Lehtokoski, A. 184
Leinonen, L. 41
Lehheriscy, S. 31, 33
Leonard, J. A. 37
Lepore, F. 185
Lessard, N. 185
Levick, Y. S. 32, 33
Levin, A. V. 35, 41
Levine, D. S. 34
Levy, R. 31, 33
Lewald, J. 184
Lewis, T. L. 35, 41
Liben, L. S. 36, 135
Lima, F. de 85
Lindell, A.K. 89
Linnankoski, I. 41
Liotti, M. 185
Lippa, Y. 69, 163

Lloyd. D. 167
Lobel, E. 105, 106
Lockman, J. J. 39
Lodesani, M. 88
Logie, R. H. 30
Longmire, S. P. 144
Loomis, J. M. 36, 37, 67, 69
Lund, F. H. 51, 52

McCarthy, R. A. 146
McClelland, J. L. 23
MacDonald, J. 28, 30
McGurk, H. 28, 30
McGlone, F. 167
Mack, A. 149
Maculoso, E. 167
Maioli, C. 106
Malach, R. 81, 169, 179, 184
Malone, P. 147
Maloney, T. 122
Marchetti, F. M. 124
Marks, L. E. 123, 126
Marmor, G. S. 36
Mason, A. 50, 179
Massaro, D. W. 146, 163
Maurer, D. 35, 41
May, M. 52
May, M. G. 146
Mayes, A. R. 44, 106
Mehta, Z. 89
Meltzoff, A. N. 28
Menon, R. S. 169
Merabet, L. 184
Meredith, M. A. 28, 32, 34
Merzenich, M. M. 101
Mesulam, M. M. 34
Metz, R. 41
Metzler, J. 24, 39
Michel, C. 41
Millar, S. 22, 23, 25, 28, 29, 30, 35, 36, 38, 39, 40, 42–44, 48, 49, 50, 54, 56, 75–77, 78, 79, 80, 82, 83, 85, 86, 87, 90, 91, 96, 99, 100, 102, 104, 107, 108, 116, 126, 134, 138, 139, 147, 149, 154, 155, 163, 172, 173, 190
Miller, G. A. 21, 30
Milner, A. D. 29
Milner, B. 87
Minami, K. 89
Mishkin, M. 43, 169, 179
Mitchell, D. E. 8

Moore, B. O. 34, 105, 106
Morris, N. 113
Morris, R. G. M. 41
Moses, F. L. 149
Moss, S. 96
Mountcastle, V. B. 12, 74
Mousty, P. 82
Munakata, Y. 23
Munsinger, H. 40
Myers, D. S. 88
Myerson, J. 116

Naatanen, R. 41
Nadel, L. 34
Nelson, B. 162.
Nettleton, N. C. 87
Neville, H. 41, 185
Newcombe, F. 89
Newcombe, N. 135
Newell, F.W. 9, 41
Newman, C. 101, 125
Newman, R. C. 37
Newport, R. 163, 164, 166
Nickerson, R. S. 75
Nicoletti, R. 162
Nicolls, M. E. R. 89
Nixon, P. D. 34, 88
Nolan, C. Y. 73, 75
Nokes, L. 50, 179
Norman, J. 23
Novikova, L. A. 9
Nyman, G. 41

O'Connor, N. 87
Ohgaki, T. 106
O'Keefe, J. 34
Ollinger, J. M. 81
Oonk, H. M. 116
Over, R. 145, 148, 159

Paavilainen, P. 41
Paillard, J. 33, 44, 69, 87, 90, 101, 106
Paré, M. 185
Pascual-Leone, A. 81, 82, 91, 168, 169, 184, 185, 186
Pashler, H. 101
Passingham, R. E. 34, 88
Patterson, J. 145
Patterson, M. Q. 74, 81
Pavlov, I. P. 11

Pearce, H.-J. 145
Peled, S. 81, 169, 179
Pellegrino, J. W. 36
Pendleton, L. R. 30
Pepper, R. 134
Peterson, L. R. 21
Peterson, M. J. 21
Perkins, G. M. 28
Piaget, J. 148
Pick, H.-L. Jr. 39.
Pietrini, P. 169, 179
Pillon, B. 31, 33
Pizzamiglio, L. 105, 106
Pochon, J. P. 31, 33
Podreka, I. 37
Poline, J. B. 31, 33
Pollack, R. H. 143, 147
Pont, S. C. 101
Poo, M.-M. 22, 23
Porac, C. 149, 156
Poranen, A. 41
Pouget, A. 22
Predebon, J. 148, 149, 154
Pressey, A. W. 149
Pressey, C. A. 149
Price, C. 81, 169
Prinzmetal, W. 122, 123
Proctor, R. W. 162, 163

Rabb, B. 163, 164, 166
Raichle, M. E. 81
Rains, G. D. 87
Rauschecker, J. P. 184
Reales, S. 30, 76, 116, 138
Recanzone, G. H. 101
Reed, C. L. 31
Reeve, T. G. 162
Reid, R. L. 123
Renowden, S. 34, 88
Revesz, G. 13, 100
Révész, G. 119, 145, 148
Ricciardi, E. 169, 179
Richardson, B. 167
Rider, E. A. 48
Rieber, C. H. 145
Riesen, A. H. 9, 41
Rieser, J. 39, 44, 48
Riggio, L. 89
Rizzolatti, G. 34, 167
Roberts, N. 44, 106
Robertson, A. 145

Robertson, I. H. 28
Rock, I. 15, 19–20, 27
Robinson, J. O. 149, 156
Röder, B. 40, 41, 185
Rösler, F. 41, 185
Roderiguez, M. 88
Rogers, B. 146
Rogers, B. J. 14
Rogers, C. M. 28
Rollag, M. D. 179
Rolls, E. T. 34, 105, 106
Rorden, C. 27
Rose, D. 101
Ross, H. 51, 52
Rouse, D. L. 51, 52.
Rudel, R. G. 28, 29, 87, 147
Rudell, A. P. 162
Rumelhart, D. E. 23
Rushworth, M. F. S. 34, 88
Ryder, K. 185

Sabate, M. 88
Sadato, N. 81, 169
Saetti, M. C. 144
Sakata, H. 32, 33
Salik, S. S. 120, 122, 146
Sanes, J. N. 106
Sathian, K. 74, 82, 184, 185
Schaeffer, A. A. 51
Schiano, D. J. 102, 149
Schicke, T. 40
Schlaegel, T. F. 36
Schlaug, G. 184
Schluter, N. D. 88
Schmidt, W. 139, 154
Schneider, W. 97
Scholey, K. A. 113
Scotti, G. 88
Scroggs, E. 120, 122, 146
Seeger, C. M. 162
Sekuler, R. 148
Semenza, C. 88
Senden, M. von 8, 41
Serrati, C. 89
Shatz, C. J. 28, 29, 41
Shepard, R. N. 24, 39
Shiffrin, R. M. 21, 97
Shorland, B. 167
Siegel, A. 49
Simon, J. R. 162
Simmons, R. W. 90, 102

Simpson, P. J. 101
Sinkonen, J. 41
Sittig, A. 37, 163, 164
Slator, R. 49
Smith, M. O. 87, 89
Smyth, M. M. 30, 113
Snook-Hill, M. M. 49
Snyder, A. Z. 81
Snyder, L. H. 22, 32, 105, 165
Spehr, K. 87
Spelt, D. K. 121
Spence, C. 30, 31, 40, 163
Spivey, K. M. J. 154
Springbett, B. M. 146
Stanford, T. R. 28.
Steffen, H. 145
Stein, B. E. 28, 30, 32, 34, 163
Stein, J. F. 32, 34, 89, 101
Steiner, M. 37
Sterr, A. 41, 185
Sterzi, R. 165
Steven, M. S. 184
Stevens, J. C. 74, 81
Strelow, E. R. 49, 55
Streri, A. 28
Stuart, G. 113
Swets, J. A. 23, 167
Summala, H. 41
Summers, C. D. 37
Suzuki, K. 159

Tam, M. 122
Tamietto, M. 36
Taylor, C. S. R. 167
Taylor, L. 87
Taylor-Clarke, M. 166, 167
Teder-Sälejärvi, W. 41, 185
Tempelton, W. B. 14–16, 28, 44, 52, 69, 121
Teuber, H.-L. 28, 29, 147
Theoret, H. 184
Thiagarajah, M. C. 74, 82
Thomson, J. A. 48
Tian, B. 184
Tinti, C. 36
Tipper, S. P. 19, 167
Titchener, E. B. 10
Tobin, M. J. 81, 96
Tolman, E. C. 18, 44
Toni, I. 144
Tooze, S. 96

Torres, F. 81, 186
Travieso, D. 145
Treisman, A. 86
Tsai, L. S. 145, 147, 154
Turnbull, O. H. 146
Turnure, C. 28
Tversky, B. 102

Umiltà, C. 162.
Ungeleider, L. G. 43, 169, 179

Vallar, G. 87, 89, 105, 106, 165
Vecchi, A. 88
Vecchi, T. 36
Veraart, C. 41
Verenka, A. 102
Victor, J. 15, 27
Villardi, K. 149
Vurpillot, E. 73, 89

Wade, A. R. 146, 159
Wade, D. T. 34, 88
Wade, N. 159
Wall, R. S. 49
Walker, J. T. 145
Wallace, J. G. 35, 146
Wallace, M. T. 28
Wallace, R. J. 162
Wallace, S. A. 90
Walsh, V. 88
Wanet-Defalque, M. 41
Wards, R. 44
Warren, D. H. 35
Warrington, E. K. 89
Wassermann, E. 81
Watt, H. J. 10
Wattam, B. 50, 179
Weber, E. H. 9, 166

Weeks, D. J. 162
Weiskrantz, L. 187
Welsh, R. A. 67
Wenderoth, P. 149, 159
Wertheimer, M. 11, 17, 28, 168
White, C. W. 148
White, S. H. 49
Wiles-Kettleman, M. 73
Willen, J. D. 147
Williams, M. 52, 67
Williams, P. A. 122
Wilson, J. A. 149, 156
Wilson, K. 145
Wilson, L. 116
Woldorff, M. G. 185
Wolfe, U. 122
Wolpert, M. 31, 163
Wong, T. S. 124, 131, 134, 148
Woodin, M. E. 104
Worchel, P. 35, 51, 52
Wu, W. H. C. 169, 179
Wuillemin, D. B. 167
Wundt, W. 10

Xing, J. 32

Yoneyama, K. 145
Youngquist, G. A. 48

Zaback, I. A. 36
Zangaladze, A. 82, 184
Zhao, Z. 36
Zeki, S. 32, 33
Zhou, R. 181
Zhou, Y. D. 185
Zipser, D. 33
Zohary, E. 81, 169, 170, 179, 184
Zoppello, M. 87

Subject index

Ability 35
   as potential 36, 50, 51
   'special' populations and small
     samples 35, 38
   standardised tests 36
Active touch (see also Touch, Haptic
   perception) 16–17, 72, 73
Attention 20, 30
   age and 'centration' 148
   involuntary orienting to the source of
     stimulation 30, 168
   selection, pre-attentive effects 86
   shifts and reduction (automation)
     with repetition 97, 148, 187
   vigilance (sustained attention)
     21
Allocentric (external, extrinsic,
   exteroceptive, environmental)
   reference cues 5, 13 (see Spatial
   processing)

Bimodal inputs in vision and touch 30,
   105, 167 (see Crossmodal effects)
Bimodal neurons and "personal space"
   33, 105
Bisection illusions and shape orientation
   15, 121 122, 131, 132, 133
   line bisection points as junctions or
     boundaries 122, 130, 135–141,
     188
Body coordinate systems (see also
   Vestibular system) 14–19
   feedback (re-afference) from
     self-produced movements 15,
     28
   gravitational cues and anti-gravity
     muscles 14, 15
   physical, anatomical, physiological
     and ecological constraints 14
Body-centred (egocentric) reference cues
   15, 40, 44, 48 (see Spatial
   processing)
   coordination with external cues in
     vision and in touch 105, 113,
     165
   the vertical (upright) and mid-body
     axes as cues to direction 14, 15
Braille and print characters (see also
   Finger movements) 73–5
   connections between contributing
     factors 86, 183
   contour perception, shape coding
     and lack of distinctive features 75,
     76

SUBJECT INDEX

long-term experience and age 74, 81, 82
Brain regions (cerebral hemispheres) 32, 33

CAPIN (convergent active processing in interrelated networks) model 42–44 (*see* Spatial processing)
Coding inputs across brief delays 21 (*see* Short-term and Working memory)
re-coding sensory inputs 30
Cognitive (central, 'higher order')
activities and conscious awareness 19–21, 24, 97, 166, 167, 187
Connectionist models and non-linear mathematics 22–23
Convergence of multisensory inputs in cerebral (spatial processing) regions 32–34, 105, 165, 169
Crossmodal (intersensory) effects 16, 27–31
at birth and with development 28
cerebral spatial processing areas and multisensory inputs 30, 32–34, 42
vision, hearing and speech 30
methods and within-subject designs 28, 189
stimulus complexity, age and verbal, non-verbal tasks 28, 29
visual capture ("dominance") and task demands 15, 28, 31
visual-haptic associations 185
Common stimulus characteristics and discrepancies 192, 193

Depth perception and vision 14, 146
Distance (length) perception in touch 93, 95
kinaesthetic, movement and location cues 90–93, 102
Dichotomies as initial distinctions between interrelated factors 9, 22, 192
Dorsal and ventral stream processing of spatial and shape cues 43, 106, 169, 179

Egocentric (body-centred) reference or orientation 15 (*see* Spatial processing)

Euclidean coordinates describe the angles of body parts relative to gravity and anti-gravity supports as directional axes 13, 14
visual and haptic space judgements as non-Euclidean 101, 102, 103
recognising and learning to apply geometric rules 7, 50, 58
Evolution and advances in biology, microbiology and genetics 8

Finger movements: shear patterns in lateral scanning and feature detection 72, 75, 79, 80, 182
Forms of spatial organisation as modality-specific 10, 13, 100
body-centred and external reference in vision and touch 105, 108, 112, 187

Gravitational cues 14 (*see* Inputs from vestibular system)

Hand movements (*see also* Laterality)
pick-up of spatial and verbal information in two-handed scanning 83–85, 182
reaching and re-afference 15, 28
feedback from vision and proprioception 105
Haptic (touch and movement)
perception 2, 12, 16, 30, 72, 73 (*see also* Hand movements and Finger movements)
explorative movements and touch in identifying objects 30, 31
haptic spatial processing 101, 102, 111, 112, 113
Haptic and visual 'spaces' as non-Euclidean 101, 102, 103
left-right bias relative to inadequate reference cues 101
Heading directions in urban environments and biasing cues 49, 182
Hearing and echo-location 49, 50
Hippocampus and short-term spatial coding and memory 34, 106
connections to the vestibular system 106

external cues and neural processing streams 106
reciprocal connections with neocortex 34

Illusions 120 (*see* Bisection, Vertical-horizontal, and Müller-Lyer illusions)
in blind and sighted people 119
as discrepancies in normally converging, informative cues 119, 120, 140
discrepant shape-based reference cues 120, 125, 141, 145, 159, 188, 189
latency differences in scanning speeds 124, 127, 150, 152,
lateral inhibition 146
size-distance constancy scaling 122, 145, 146
stimulus range (context) effects 123, 126
eye and hand movements as factors 148
Imagery as visual 13, 24, 36–37, 39
as mental representations, strategies or heuristics 10, 37, 39
as covert movement rehearsal 30, 102, 116
Implicit awareness of directions 18
Information as defined in communication systems 20, 22
as detection of invariant features 19
imbalance of inputs in the absence of sight 35, 37
"redundant" information and benefits of dual inputs 21, 30, 31, 40

Kinaesthetic and proprioceptive inputs in vision and touch 12, 102, 116, 165

Landmark cues in touch and vision and order effects 49, 50, 113, 172
Laterality (hand use) and the cerebral hemispheres 72, 87, 88, 89
cerebral comissurotomy and hemi-neglect 87, 89
cross-lateral effects 91, 92, 93–6
Locomotion in large-scale space 48–53 (*see* Veering)

inferred directions and task conditions 49, 182

Memory stores (*see also* Short-term memory and Working memory) 21, 49, 187, 191
longer-term memory, effects of experience and prior information 190
pre-and post-synaptic activities 21, 22
Model of statistically optimal integration of available inputs 163, 164, 180, 193
Modalities as multisensory perceptual systems 12, 16–19
Molyneux's question 8, 9
Modality-specific effects and spatial reference cues 27, 100, 187, 188
Müller-Lyer illusions 143
convergent/divergent fin differences 143, 151, 153
discrepant length cues from constituent features, and external cues 154
as endpoint confusions 149, 152
latencies of eye and hand movements 148, 149
practice, repetition and perceptual learning 146–148
effects of reference instructions in touch and vision 156–158, 188
Movement control and reference cues 33, 90–93, 95, 96, 169
motor, pre-motor areas and cerebellum 32, 33

Networks as metaphors for interrelations in processing inputs spatially for perception and action 27, 42, 86, 187, 190, 193
Neural networks connecting distributed cortical and subcortical brain regions
involved in different spatial tasks 32–34, 88, 105, 106, 193

Occam's Razor 181
Occipital (visual) cortex 31
activation of visual areas by touch 81, 169, 184

232 SUBJECT INDEX

Oblique orientation and reference information 39, 102, 103, 165
Object/shape-based coding (*see also* Dorsal and ventral stream distinction, and Classification of reference cues) 43, 106, 169, 179

Parietal cortex regions and spatial processing 32, 33, 105, 106
    connections with pre-frontal cortex and subcortical regions 34
    bimodal neurons in anterior parietal areas 185
Pathways for cutaneous and somaesthetic and proprioceptive inputs 31
Plasticity of neural connections 8, 28, 41, 184, 185
    activation of cell populations and task effects 102
    activation of visual areas by touch 41, 81, 182, 184
    in adults 184
    compensatory stimulation with sensory deprivation 15, 41, 42, 184–186
    with deprivation of sight from birth 8, 9, 41, 184
    during development 41, 184, 185
    molecular changes 41
    reorganisation in somatosensory receptive fields 167
    temporary deprivation of a sensory input 184
Posture and proprioceptive cues in touch and vision 16, 164, 165
Prefrontal cortex, problem solving and control processes 34
    fronto-parietal network in spatial coding 106
Psychophysical methods 23, 24, 167
    methods for recognition and recall 21
    stimulus range effects 123, 126, 128

Rationalist/empiricist dichotomies 8, 9, 13
    "innate/learned" distinction as a misleading dichotomy 9, 12
Reference "frames": shorthand for the organism's activity in integrating diverse inputs in (multiple) reference relations 3, 19, 42–44 (*see* Spatial processing)
Reference cues broadly classified by the origin of inputs 3, 19, 44, 105
    as allocentric: objects or scenes perceived relative to external cues 44, 106
    as egocentric: objects perceived relative to body-centred cues 40, 43, 44, 106
    as object/shape-based: relation of features within shapes (*see also* Illusions) 44
Relational variables and integration of components 19
Rotation of layouts as a test of body-centred spatial coding 24, 25, 38, 50, 110

Short-term memory (see also Working memory) 21, 29, 34, 36, 116
    covert (mental) strategies inferred from performance 9
    interference methods and paradigms 21, 24, 25, 27
    limited capacity 20, 21
    memory spans for serial tactual items 29, 30
    movement spans and rehearsal 30, 102, 116
    visual short-term memory 29, 30 (*see* Visuo-spatial sketchpad)
    rehearsal 22 (*see* Working memory)
Space as an innate concept 7, 8
    as *a priori* intuition 8
    as empirical (associative) construction 8
Spatial processing (criteria) 3, 42–44, 190
    processing diverse inputs as reference (relational) cues for the location, distance and direction of objects in perception and action 2, 3, 42–44, 92, 95
Specialised senses, improved detection, and benefits of "redundant" information 31
    integrative processing for consistent perception and action 31, 106

Stimulus–response compatibility 161, 162–3
"Simon" effect and spatial reference 162–163
   irrelevant spatial cues and task demands 161–162, 190
Stimulus specificity, modularity and integrative activities 106, 190, 193

Tactile (tactual) acuity (*see also* Threshold measures) 9, 74, 82, 166, 168, 177
   age and braille experience 74, 81, 82
   improvement with magnified vision 166, 168
Tactually sensitive cells in occipital cortex 81
Task effects (*see also* Stimulus–response compatibility) 24
   influence of 'irrelevant' task conditions 162
Theories of perception 10–20
   detection of invariant, ecologically valid information 18
   direct amodal detection of relations between surfaces 18–19, 192
   direct/indirect controversy, methods of testing and task variables 16–20, 23
   Gestalt school laws and global shapes 11, 73
   as hypothesis testing 19
   phenomenology as a method 11
   sensory "atomism" and "unconscious inference" 9, 10, 11, 44
Threshold measures 9, 23, 167
   two-point threshold 9, 166, 167, 168,
"Top-down", "bottom-up" and interactive processing 96, 97, 115, 159, 191
Touch (*see also* Haptic perception) 12–13, 16
   as a multisensory system 12, 13, 16
   discrimination and distinctive features 73, 74, 149, 174, 180
   lemniscal and spinothalamic pathways 30, 31
   passive touch 12, 89
   perception of 3-D objects 31
   skin (cutaneous) receptors 12, 31, 74
   somatosensory regions of the brain 32
   texture perception 31

Veering from the straight-ahead 51
   body-structure, laterality, and sounds 52, 53,
   bias from isolated noise and posture cues 62–67
   blind and sighted pedestrians 53
   stride and walking speed 63, 66–68
Vertical-horizontal (orientation-specific) illusion 140, 141, 158, 188
   as eliptical field (boundary) effects in vision 122, 123
   as radial/tangential movement differences 124–126
   as movements bounded by the elbow or shoulder joint, respectively 130, 131
Vestibular system and sensing gravitational cues 14, 17, 48
Vision and spatial coding (*see also* Plasticity of neural connections) 17, 36, 39, 40, 49, 103, 190
   development, experience, stimulation and information 34–41
   deprivation of sight from birth, and spatial cues 38
   effects of experience and practice 39
   peripheral vision, tunnel vision, and diffuse light perception 170–178
   spatial and non-spatial vision 177, 190

Williams syndrome and vision (*see also* Parietal-prefrontal connections) 50
Working memory 22, 29–30, 116
   articulatory/phonological coding and rehearsal loop 29, 30
   episodic buffer and dimensional integration 30, 116
   limited capacity and central executive control 20, 21, 30, 116
   reciprocal connections 116
   visuo-spatial sketchpad 22, 30, 103, 116